Doing Global Science

DOING GLOBAL SCIENCE

A Guide to Responsible Conduct in the Global Research Enterprise

INTERACADEMY PARTNERSHIP

PRINCETON UNIVERSITY

Princeton and Oxford

The illustrations contained in the guide are the copyrighted work of
S. Harris, and are used with permission (www.sciencecartoonsplus.com).

Library of Congress Cataloging-in-Publication Data

Names: IAP—the Global Network of Science Academies, author.
Title: Doing global science : a guide to responsible conduct in the
global research enterprise / InterAcademy Partnership/IAP—The Global
Network of Science Academies, Indira Nath, Ernst-Ludwig Winnacker,
Renfrew Christie, Pieter Drenth, Paula Kivimaa, LI Zhenzhen, Jose A.
Lozano, and Barbara Schaal ; project staff, Tom Arrison, Anne Muller,
Steve Olson, Lida Anestidou, Nina Boston, and Patricia Cabezas.
Description: Princeton, New Jersey ; Oxford : Princeton University
Press, [2016] | ?2016 | Includes bibliographical references and index.
Identifiers: LCCN 2015045281| ISBN 9780691170756 (hardcover)
| ISBN 0691170754 (hardcover)
Subjects: LCSH: Research. | Science.
Classification: LCC Q180.A1 I23 2016 | DDC 174/.95—dc23
LC record available at http://lccn.loc.gov/2015045281

British Library Cataloging-in-Publication Data is available

This book has been composed in Linux Libertine and Raleway

Printed on acid-free paper. ∞

Printed in the United States of America

1 3 5 7 9 10 8 6 4 2

Contents

Foreword

A global research enterprise has emerged over the past several decades as significantly more research is being performed in a growing number of countries by a larger cohort of researchers. International and interdisciplinary linkages in research are expanding rapidly, with more papers being coauthored by investigators based in different countries and representing different fields. More researchers are crossing borders for education and training. The new knowledge produced by the global research enterprise promises to expand our understanding of the natural world and to accelerate progress in meeting humanity's needs in areas such as health, the environment, and economic development.

However, irresponsible behavior and poor practices pose threats to the global research enterprise, could impair its effective functioning, and could even damage the broader credibility of science. High-profile cases of data fabrication and other irresponsible behavior continue to appear around the world. Issues related to journal article retractions and the reproducibility of research results are attracting greater attention.

Opportunities for researchers to contribute to society are also expanding rapidly. Scientists are increasingly called upon to demonstrate vigilance aimed at preventing the deliberate misuse of research in the life sciences and other fields and to contribute to society as policy advisors and as communicators of scientific ideas and findings to the broad public.

In response to these trends, the world's national scientific academies, working through the InterAcademy Partnership, launched a project on research integrity in 2011. The first product of the activity was *Responsible Conduct in the Global Research Enterprise: A Policy Report*, published in 2012. That report describes basic values that underlie research and puts forward principles and guidelines for all participants in the research enterprise.

This publication was developed to assist students, individual researchers, universities and other research organizations, public and private research sponsors, journals, societies, and policy makers as they work to foster research integrity and secure the foundations of responsible conduct. *Doing Global Science: A Guide to Responsible Conduct in the Global Research Enterprise* is a concise, engaging guide to responsible research behavior. It is written from a global perspective and addresses a range of traditional and emerging issues related to scientific responsibility, using examples from various disciplines. It can be used in educational settings, by supervisors in training settings, and by individuals.

We are grateful for the hard work of the Committee on Research Integrity, which developed this guide, particularly committee cochairs Ernst-Ludwig Winnacker and Indira Nath. We also appreciate the work of the independent set of experts who peer-reviewed the draft according to IAP for Research Procedures and to the monitor who oversaw the review process.

We hope that *Doing Global Science* is widely used and expect that it will contribute to the health and effectiveness of the global research enterprise.

Robbert H. DIJKGRAAF
President, InterAcademy Partnership
Cochair, IAP for Research
Director and Leon Levy Professor
Institute for Advanced Study, Princeton, New Jersey, USA
Former President, Royal Netherlands Academy of Arts and Sciences

Mohamed H. A. HASSAN
President, InterAcademy Partnership
Cochair, IAP for Science
Chair, Council, United Nations University (UNU)
Former President, African Academy of Sciences

Preface

The world's science academies set and maintain standards of research integrity and scientific responsibility and are taking a leadership role in addressing issues related to responsible conduct. National science academies in a range of countries have issued policy recommendations, educational materials, and statements aimed at improving the environment for research integrity in their own countries. In addition, regional networks of academies have produced reports or hosted workshops and conferences on these topics. In some countries, academies play a direct role in ensuring responsible conduct and addressing irresponsible behavior through their own research and educational activities or through participation in national oversight bodies.

The research integrity project was launched in 2011 by two member networks of the InterAcademy Partership (IAP): IAP for Research (known up to now as the InterAcademy Council) and IAP for Science (known up to now as IAP—The Global Network of Science Academies). The first product of this project, *Responsible Conduct in the Global Research Enterprise: A Policy Report* (2012), clarified the primary values of responsible science, and recommended actionable steps to key stakeholders around the world. The report has been widely used, for example, as a key background document by the Global Research Council.

Doing Global Science: A Guide to Responsible Conduct in the Global Research Enterprise is the second product of the project. The project terms of reference state that IAP for

Research "will develop international educational materials for individual scientists, educators, and institutional managers, addressing principles and guidelines for scientific responsibility, including scientific ethics, integrity, and responsibility for avoidance of misuse of science. The products will have use throughout the global science community" (IAC-IAP 2012).

Committee on Research Integrity

This guide was developed by an international committee appointed by the IAP for Research Board: Indira NATH (Cochair, India), Ernst-Ludwig WINNACKER (Cochair, Germany), Renfrew CHRISTIE (South Africa), Pieter DRENTH (The Netherlands), Paula KIVIMAA (Finland), LI Zhenzhen (China), José A. LOZANO (Colombia), and Barbara SCHAAL (USA). A complete roster and biographical sketches of the committee members are included at the end of the guide.

The Committee on Research Integrity met a number of times in person, held conference calls to develop and review drafts, and responded to the comments of external reviewers in finalizing the guide.

The Review Process

This guide was externally reviewed in draft form by experts chosen for their diverse perspectives and technical knowledge, in accordance with procedures approved by the IAP for Research Board. The purpose of this independent review was to provide critical comments that would help produce a sound report that meets the IAP for Research standards for objectivity, evidence, and responsiveness to the study charge.

The review procedure and draft manuscript remain confidential to protect the integrity of the deliberative process. Although the reviewers provided constructive comments and suggestions, they were not asked to endorse the conclusions and recommendations, nor did they see the final draft of the report before its release.

Reviewers of the Report

IAP for Research thanks the following individuals for their review of this report:

Nicole FÖGER, Head of the Administrative Office, Austrian Agency for Research Integrity, Vienna

Matthew FREEMAN, Professor of Pathology and Head, Sir William Dunn School of Pathology, University of Oxford

Michele GARFINKEL, Manager, Science Policy Program, European Molecular Biology Organization

Alastair HAY, Professor of Environmental Toxicology, University of Leeds

Sergio PASTRANA, Executive Director and Foreign Secretary, Academy of Sciences of Cuba, and member of the IAP Biosecurity Working Group

Bernd PULVERER, Chief Editor, European Molecular Biology Organization

Sameh H. SOROR, Assistant Professor of Pharmacy, Helwan University, Cairo, and Cochair, Global Young Academy

Monitor of the Review Process

A review monitor was responsible for ascertaining that the independent examination of this report was carried out in accordance with IAP for Research procedures and that review comments were carefully considered.

IAP for Research thanks the following for his participation as monitor in the review process:

Yves QUÉRÉ, Former President, Académie des sciences (France), and Former Cochair, IAP—The Global Network of Science Academies

Financial Support

Financial support to undertake the project was provided by the U.S. Department of State, the U.S. National Academy of Sciences, and IAP for Science.

Doing Global Science

1

RESPONSIBLE CONDUCT OF RESEARCH AND THE GLOBAL CONTEXT

An Overview

"JUST CHECKING."

Scientific research is one of the great adventures of our time. Researchers are members of a global community that is producing new knowledge at an unprecedented rate. This new knowledge is transforming society by contributing to the development of new technologies and by changing how we think about the natural world, ourselves, and our institutions.

The growth and accelerating pace of scientific discovery has made the twenty-first century an exciting time to be a researcher. Large international teams are working on problems that were impossible to solve in the past, such as the annotation of the human genome, the search for dark matter, or the analysis of "big data" derived from social media. New fields of research are opening up at the intersection of traditional disciplines, such as nanobiology and neuroeconomics (Glimcher 2003; Nussinov and Alemán 2006). Researchers are generating knowledge that could fundamentally alter agriculture, energy production, environmental protection, communications, and many other aspects of human life. Our future on this planet will depend to a considerable extent on the products of research.

Like the rest of society, the research enterprise has been undergoing momentous changes. Information technology is revolutionizing how research is done and how researchers interact with each other. Most researchers work not just on individual tasks but as parts of research teams that include people with many different backgrounds and perspectives and may be international in scope. Governments around the world, recognizing the critical role of research in improving the well-being of their citizens, are increasing their support for science and engineering. As a result, millions more scientists and engineers are working today than was the case just two decades ago (NSB 2012).

The changes going on within research have created challenges. Team research can create conflict as well as opportunity. The rapid expansion of the research community may disrupt the transmission of traditions and ethical principles to new researchers. Increased competition for resources may intensify the pressures faced by researchers, including young investigators, to publish more papers and to publish in the most prestigious journals. Technology-enabled tools such as blogs and social media increase the speed of scientific communication but may also contribute to eroding collegiality or facilitate the spread of unreliable information. Many researchers travel to countries where they may encounter different research practices than they are used to, or they may become involved in interdisciplinary research that is unlike research they have done before. The research landscape has become more diverse, more interconnected, faster paced, and more complex than ever.

Throughout the history of research, young and early career researchers have learned about standards of conduct by working with more experienced researchers. This process of learning by doing will continue to be essential in the training of future generations of researchers. However, new researchers can benefit from having a readily accessible and compact source of guidance—guidance that more-experienced researchers need to review and follow as well. All researchers can benefit from a better understanding of changes in the research landscape and their possible impacts.

In 2008, the International Council for Science (ICSU) published a booklet providing guidance about the responsibilities and freedom of researchers to maximize the benefits of science for society. One year later, the educational guide *On Being a Scientist: A Guide to Responsible Conduct in Research* (NAS-NAE-IOM 2009) was published. In 2012, the InterAcademy Council (IAC) and IAP—The Global Network of Science Academies—published *Responsible Conduct in the Global Research Enterprise: A Policy Report*, which describes the values of research and how those values should guide the conduct of research. This 2012 report acknowledged that different disciplines and countries have varying research traditions and cultures. But it argued that the fundamental values of research transcend disciplinary or national boundaries and form the basis for principles of conduct that govern all research.

The educational guide you are reading now is an adaptation and expansion of the earlier policy report and was written by the same committee. It includes much of the same content and in some cases even the same language; text from the recommendations of the report is printed in boldface type when it appears in this publication. However, this publication has a different goal than previous documents. It has been written as a practical guide to conduct in a research environment that is being transformed by globalization, interdisciplinary research projects, team science, and information technologies. It addresses both long-standing issues in the responsible conduct of research and emerging issues. It is aimed not only at new researchers but at more-experienced researchers and research administrators, funders, and policymakers, all of whom are caught up in the broad trends that are reshaping the research enterprise.

This guide provides an overview of which research behaviors are responsible and to be embraced and which are irresponsible and to be avoided. It uses specific examples from a variety of areas to provide guidance relevant to researchers in all fields. The organization of the guide parallels the research process. The even-numbered chapters follow the process of research, from planning and preparing to undertake research (chapter 2) to carrying out research (chapter 4), to preventing and addressing irresponsible research practices (chapter 6), to reporting research results (chapter 8), to communicating with policymakers and the public (chapter 10). The odd-numbered chapters discuss broader issues associated with performing research: the researcher's responsibilities to try to prevent the misuse of research and related technology (chapter 3), the researcher's responsibilities to society in planning and carrying out research (chapter 5), aligning incentives with responsible research (chapter 7), and the benefits and challenges of international collaboration (chapter 9). The references and additional resources do not represent an exhaustive bibliographic source, but they provide the reader with further material about the topics covered in each chapter.

Two types of boxes accompany the text. Focus boxes illustrate the issues discussed in each chapter. Discussion scenarios describe hypothetical situations and related questions to foster debate.

A key premise of this project is that prevention is better than cure—that more and better efforts to educate and train researchers about the importance of adhering to high standards and good practices will speed the advance of knowledge and increase the positive impacts of research. Many publications are available that describe responsible conduct in science. What sets this guide apart is its emphasis on internationally harmonized standards in a rapidly changing global research environment. Some of these standards are still in flux and are not yet universally observed. But every researcher has a responsibility to contribute to the development and dissemination of these standards, just as every researcher has an obligation to maintain the integrity of research. Societies around the world have placed their trust in scientific research to generate knowledge for its own sake and to understand and solve major problems. To maintain this trust, everyone involved with the research enterprise must help ensure that research is conducted responsibly.

Terminology and Definitions in This Guide

Research

This report treats research as encompassing many forms of disciplined human thought, including the natural sciences, the social and behavioral sciences, and the humanities. Research thus includes the generation of new knowledge in fields traditionally recognized as the sciences, whether theoretical, experimental, or computational, and in other areas grounded in the rational analysis of empirical evidence.

Irresponsible Conduct, Practices, or Behavior in Research

In this report, all unethical and harmful behaviors by researchers that relate to the conduct of research are referred to as irresponsible research practices, behavior, or conduct. The report refers to ethical and desirable research-related behaviors as responsible research practices or responsible research conduct.

Misconduct and Fraud

Different countries define research misconduct and research fraud to include serious categories of irresponsible research practices such as fabrication or falsification of data or plagiarism. Some countries include as misconduct or fraud such behaviors as obstructing an investigation into research misconduct or retaliating against a whistle-blower.

Bias

For the purposes of this report, a bias is a tendency or inclination on the part of a researcher or research group that introduces systematic error into the research process and damages the validity of the resulting work. Biases can affect research design, data collection and interpretation, or the reporting of results. While biases may be difficult or impossible to eliminate completely, steps can be taken to identify and minimize the most serious potential sources of bias.

Conflict of Interest

A researcher is considered to have a conflict of interest when financial, personal, or other considerations have the potential to compromise judgment or objectivity. Research sponsors and research institutions often require researchers to disclose possible conflicts of interest and may institute additional oversight procedures or restrict involvement of the conflicted researcher in the work.

Principal Investigator (PI)

This term refers to the senior researcher in a laboratory or research group. PIs are often the primary supervisors of graduate students and postdoctoral fellows and are responsible for tasks such as submitting proposals and complying with research-related regulations.

Possible Ways of Using this Guide

This guide can be used in many different ways. It can be read by individuals, discussed in groups, or taught in seminars or classes. It can form the basis for an online seminar or discussion involving larger or more-distributed groups. A research group or institution could use it to supplement existing codes of conduct. Or it could be used to develop a code of conduct for a specific research field or institution. It is short enough to cover in a single workshop or can be combined with other materials in a broader treatment of responsible research.

The committee has sought to keep the guide concise. A growing wealth of excellent materials on research integrity and scientific responsibility is available, and the "References and Additional Resources" sections at the end of each chapter provide the necessary information about accessing such materials.

The discussion scenarios in this guide have been designed to foster examination of difficult questions. They do not have simple answers or easy solutions. One way to use these discussion scenarios would be to assign individuals or groups to present and defend contrasting viewpoints. Discussants could identify affected parties—whether individuals, institutions, research fields, or society—and the interests each party has in the situation. They then could explore possible actions and the consequences of each action. Discussants also could be encouraged to act out the roles of parties with conflicting interests to explore more deeply the tradeoffs and uncertainties associated with possible actions.

2

PLANNING AND PREPARING FOR RESEARCH

"FIND OUT WHO SET UP THIS EXPERIMENT. IT SEEMS THAT HALF OF THE PATIENTS WERE GIVEN A PLACEBO, AND THE OTHER HALF WERE GIVEN A DIFFERENT PLACEBO."

Developing research ideas and a research plan are among the most exciting parts of the research process. It is a process that combines creativity, collaboration, judgment, and experience. It also involves the fundamental values of research and the principles of responsible conduct derived from those values. Good mentors can be extremely valuable to younger researchers with less experience in planning research.

Many issues of responsible conduct arise during the development of research ideas and plans. The sources of research ideas need to be acknowledged. Research proposals and plans are often peer reviewed before work begins. Interdisciplinary research is becoming more important, which has implications for peer review. This chapter describes issues to consider in planning and preparing for research.

The Values of Research

Responsible conduct in research is based on fundamental human values that apply in many other domains of human life. But the basic values that underlie research have specific implications for the conduct of research. The application of those values in developing a research plan and in successive phases of research produces distinct principles that can guide the actions of researchers and often dictate particular practices, such as full and accurate reporting of research results.

This guide is based on seven fundamental values of research:

- Honesty
- Fairness
- Objectivity
- Reliability
- Skepticism
- Accountability
- Openness

The list draws on a number of other guidelines and reports on research integrity from recent years (3rd WCRI 2013; AG-NHMRC-UA 2007; CAS 2007; CCA 2010; DFG 2013; ESF 2010; ESF-ALLEA 2011; GBAU 2004; IAS 2005; ICB 2010; JANU-JAPU-FJPCUA-SCJ 2014; NAS-NAE-IOM 2009; NRC-IOM 2002; RIA 2010; SCJ 2006; Steneck 2007).

In research, being *honest* implies doing research and communicating about results and their possible applications fully and without deception.

Being *fair* means treating others with respect and without bias, whether in citing a colleague's ideas in a paper or mentoring a student in the proper conduct of research. In research—as in life—scientists and scholars should treat others as they hope and expect to be treated in return.

Objectivity implies that researchers try to look beyond their own preconceptions and biases to the observation and registration of facts and to the empirical evidence that justifies conclusions. Researchers cannot totally eliminate the influence of their own perspectives from their work, but they can strive to be as objective as possible.

Research communities over many years have developed methods to enhance the *reliability* of the results they obtain, and researchers have an obligation to adhere to these methods or demonstrate that their alternative approach produces equally trustworthy results.

An allegiance to empirical evidence requires that researchers maintain a degree of *skepticism* toward research results and conclusions so that results and explanations are continually reexamined and improved.

Researchers are *accountable* to other researchers, to the broader society, and to nature. If challenged, they cannot appeal to authority but must demonstrate that their results or statements can be justified.

Finally, researchers need to be *open* with others for research to progress. All researchers deserve to work independently as they balance the competing considerations of whether their hypotheses are supported or not. But they ultimately need to convey to others their conclusions and the evidence and reasoning on which their conclusions are based so that those conclusions can be examined and extended. For the empirical and experimental sciences, this requires

careful storage of data and making the data and other information underlying reported results publicly available.

The primacy of these seven values explains why trust
is a fundamental characteristic of the research enterprise.
Researchers expect that their colleagues will act in accord
with these values. While there are examples of situations
in which applying research values is not straightforward—
such as in psychological research that may involve deception of research subjects during an experiment—these cases
are highly unusual, and deviations need to be reported in
full. When a researcher fails to adhere to one of the values of research, that person's trustworthiness is diminished
among other researchers. In addition, the public's trust in
research can be damaged, with harmful effects on the entire
research community.

The Importance of Mentoring

For most young and early career researchers, mentors will
be fundamental sources of advice and guidance in developing a research plan as well as in other areas. Indeed, the
most valuable lessons a researcher learns are more likely
to come from mentors or peers than from textbooks. A faculty advisor or supervisor can serve as a mentor, but strong
mentoring generally implies a closer and more personal
connection. Mentors help their trainees and students understand and contribute to the scientific enterprise. These
trainees and students may, in turn, carry on aspects of a
mentor's work. As one report states, "a good mentor seeks to
help a student optimize an educational experience, to assist
the student's socialization into a disciplinary culture, and
to help the student find suitable employment. These obligations can extend well beyond formal schooling and continue
into or through the student's career" (NAS-NAE-IOM 1997).

Mentors have a special responsibility to help new researchers master the ethical dimensions of research. While

guides like this one can describe the general values and principles on which research is based, the application of those values and principles can vary by discipline and research tradition. Mentors can explain how specific practices reflect and reinforce the responsible conduct of research and ensure that this understanding is passed from one generation of researchers to the next. Transmitting values and practices to international students and postdocs can involve special challenges, such as overcoming language and cultural barriers. Given the growing importance of cross-border education and research, senior researchers and institutions need to ensure that all members of the team and all coauthors of resulting papers understand research values and their practical application.

Because mentoring is a one-to-one relationship, the quality of mentoring can vary among trainees or students. A possible consequence is that some fraction of young researchers will not be as effectively trained as they should be, and in some cases conflict might arise (see box 2-1). A recent report from the Global Young Academy revealed that young scientists from across the globe agree that the lack of proper mentoring can be a critical obstacle to achieving success in their careers (Friesenhahn and Beaudry 2014). A good mentor should provide career guidance, exchange ideas, discuss results, enhance networking, and promote a healthy balance between personal and professional interests. They are role models and linchpins in maintaining and fostering responsible conduct in the research environment.

All members of the research system need to take steps to ensure that every new researcher receives effective mentoring. Research institutions have special responsibilities. Faculty members at research institutions are rewarded primarily for their publication record and other aspects of performance unrelated to mentoring. Mentoring related to responsible conduct may be seen by some faculty as an "unfunded mandate." To provide stronger incentives for faculty

**Box 2-1. Discussion Scenario:
An Inattentive Supervisor**

You are a doctoral student in ecology and evolutionary biology. You have been working on a grant proposal to spend three weeks in Panama doing fieldwork. If funded, you will be able to collect some key samples to complete the last chapter of your PhD dissertation. Your supervisor was appointed chair of the department last year, and since then you have been struggling to meet with her and discuss your progression. The grant proposal is due in less than a week, and although your supervisor has had the draft for a month, you still have not heard back from her.

You have been feeling uneasy during the last year because your supervisor has been contacting you at the last minute, canceling meetings, or not answering emails. You really want to avoid being in that position again. In addition, you would like to request more time and support from her in the last stages of your research project.

What can you do to get your grant proposal reviewed on time? How should you approach your supervisor to talk about your discomfort? What practices should the department promote to prevent these conflicts?

members to be more effective mentors, institutions should consider mentorship during hiring and promotion processes and should perhaps develop more direct financial rewards for mentors. Some institutions are exploring new approaches to facilitating effective mentorship for all graduate students, such as training for mentors and supplementary career development programs (NRMN 2015). Ultimately, institutions have the responsibility to ensure that students and postdoctoral fellows are aware of their rights as well as the rules and practices governing research.

Broader efforts to encourage effective mentoring can also have a positive impact. For example, the journal *Nature* has instituted an annual program of prizes for mentorship in science. In addition, as part of its scholarship program for doctoral students in mathematics, physics, chemistry, computer science, and engineering, the Deutsche Telekom Foundation supports mentorship for scholarship recipients (National Pact for Women in MINT Careers 2013).

Formulating, Acknowledging, and Protecting Research Ideas

Producing a research plan requires thought and discussion. Drawing on their accumulated knowledge and personal creativity, researchers sift through what is known in deciding what to do next. Some ideas arrive in a flash of inspiration; Einstein believed that "imagination is more important than knowledge" (Viereck 1929). Other ideas are slowly hammered out through discussion and revision. But the result is a valuable product of the research community—a question that calls out to be answered.

One aspect of excellent research is the importance of the questions addressed. Researchers and research teams should ask whether their planned research will make a significant contribution to advancing knowledge. Judgments about the value and importance of research questions will depend on the time and the field. Both revolutionary and incremental science are valuable and necessary.

The source of a research plan is often not clear in retrospect. But ideas are the lifeblood of research, and people expect and deserve to be acknowledged for their contributions to a research plan (see box 2-2). Also, researchers receive recognition for their contributions to the collective work of the research community, and this reward system is a powerful and useful motivating force in research. Thus, **researchers have a responsibility to acknowledge the**

**Box 2-2. Discussion Scenario:
Preventing Plagiarism**

You are a postdoctoral fellow in a research group. A fellow postdoc who is relatively new to the group and whose native language is different than the one used in the lab is preparing a funding proposal with the PI for a government agency and comes to you for help with a draft. In the course of the conversation, you find out that the PI has given the postdoc several previously submitted proposals from the lab to use as examples but has not given clear guidance on how to use them. The postdoc's draft proposal contains original text describing the research to be performed, which requires some editing. The draft also contains several large blocks of text that were simply copied and pasted from the example proposals. The agency's submission deadline is coming up quickly and you are preparing your own proposal for funding.

Do you consider this conduct plagiarism? What would you tell the other postdoc? What steps should the principal investigator take to ensure that students and postdocs are familiar with appropriate practices when using previous work?

source of ideas and provide credit when using others' ideas.

Before research ideas are discussed in public, they generally remain privileged as researchers, individually and collectively, work out the difficult problems associated with gaining reliable knowledge. Sometimes this phase of the research process can involve a delicate balancing act. Ideas get better when they are discussed with others. But ideas can be difficult to protect if they are discussed widely, especially as electronic communications continue to facilitate the flow of information. **Until a research idea is publicly and ethically disseminated, researchers have an obligation to**

protect privileged information about planned or proposed research.

Peer Review of Research Proposals

Before resources are committed to proposed research, a written proposal typically is evaluated by other researchers and research managers. In private industry, this evaluation generally occurs within a company or within an industrial consortium. For academic or government research, peer review—also known as merit review—is the method most often used to judge ideas. It involves the evaluation of a research proposal by qualified professionals, such as researchers working on questions closely related to those addressed by the proposal. For example, a funding agency may set up a review panel of experts in order to evaluate proposals responding to a specific solicitation or request for proposals. **Review of research proposals is an essential component of the research enterprise and a basic obligation of researchers** (see chapter 8 for further discussion of peer review).

Peer review can determine the allocation of many types of resources besides research funding, including prizes, employment, promotion, or the use of equipment or facilities. Alternatives to the peer review of proposals exist and in some cases are widely used. For example, program managers may be tasked with allocating funding. Other alternatives, such as supporting investigators rather than projects or funding research teams to achieve particular goals, may still involve expert review at some stage.

While peer review is an important tool for scientific decision-making, it has limitations. Research proposals and the review process often undervalue a critical characteristic of proposed research—the uncertainties that are likely to accompany the knowledge that will be generated. Investigators need to be honest about both the anticipated benefits of proposed research—generally expressed as a justification for funding the proposal in terms of the funder's review criteria—and the inevitable limits of that research. When this information is not included in a research proposal, funders and reviewers should make an effort to obtain it. A more complete picture of the research to be done contributes to better funding decisions and to more realistic expectations of the value of research results.

Peer review can be abused by both researchers and reviewers. When researchers ask multiple funders to support a project, they have a duty to notify these funders, since failure to do so is deceptive and could put a strain on the resources available for reviewing. Researchers submitting proposals may give too little credit to others or even plagiarize in a misguided effort to boost their own credentials or because they do not understand or ignore the conventions for providing credit. Proposed research may not adhere to regulatory norms in such areas as research involving humans, animals, or the environment (discussed in chapter 5). Reviewers may inappropriately use language or ideas from the proposals they review in their own work, or they may allow conflicts of interest to influence their recommendations. Researchers and reviewers need to be diligent in adhering to correct practices. In addition, all members in a research team should be aware of appropriate scientific writing practices in order to avoid plagiarism, such as using quotation marks to enclose verbatim text taken from source material and providing citations for material that is paraphrased or summarized (Roig 2006).

Review tends to be a conservative process and may be subject to bias (Johnson 2008). The use of specific, well-defined measures and multiple dimensions of quality can help overcome biases. But individual reviewers also have an obligation to reflect on biases they may have and seek to minimize their effects. Providing a list of excluded referees, as allowed by the *Physical Review Letters*, can help to unmask or avoid conflicts of interest (PRL 2014). **When** someone asked to review a research proposal has a conflict of interest or bias that could be seen as influencing the review, that person has an obligation to describe the conflict of interest or bias to the individuals or organization requesting the review and may need to withdraw from the process.

Interdisciplinary and International Research

Although most research is organized according to discipline or field, the history of science is replete with examples of unexpected advances occurring as a result of knowledge generated in unrelated areas. For example, discovery of the magnetic properties of atoms by nuclear physicists in the 1940s and 1950s contributed significantly to the development of magnetic imaging for medical applications several decades later (KVA 2012). This cross-fertilization of fields poses a challenge to narrowly based forms of peer review.

The recent growth of interdisciplinary, international, and data-intensive research has also complicated the peer review of research proposals. Interdisciplinary proposals can range beyond the expertise of any one individual. In these cases, review panels should include people from different disciplines so that the group as a whole has a working knowledge of the disciplines encompassed by a proposal. Alternatively, a proposal may be reviewed by more than one disciplinary group, though the separate reviews then need to be combined in some way to yield an overall assessment.

Researchers with different perspectives and backgrounds can strengthen the review process (see box 2-3). Thus, review panels that include researchers from different disciplines and different countries can improve the allocation of resources, despite the complications this diversity may entail. International research agencies tend to use reviewers from different countries to overcome local differences. In some countries, the number of individuals within a

Box 2-3. Focus: International Reviewers and Research Proposals

As the advantages of peer review in funding decisions have become more widely recognized, the use of international reviewers for national solicitations has increased. For example, DFG (Deutsche Forschungsgemeinschaft, or the German Research Foundation) reported that 22 percent of reviewers consulted in 2007 were based outside Germany (Van Noorden 2009). An Irish official has reported that the systematic use of international review panels, often with nonvoting Irish chairs to ensure that correct procedures are followed, has helped upgrade Irish research (O'Carroll 2009). In 2009, the Italian government, which has traditionally funded biomedical research through appropriations to institutions, outsourced the review of one thousand proposals to the U.S. National Institutes of Health (Van Noorden 2009). These practices, which are increasingly common, can make the review process more transparent and less prone to bias.

discipline who are qualified to review a proposal may be small, requiring the use of international reviewers. **Where possible, research sponsors should use a broad range of reviewers, including international reviewers.**

In some cases, the list of reviewers for proposed research is expanded beyond researchers familiar with the proposed work to include researchers in distantly related fields and individuals not involved in research. This form of review can introduce a wider range of considerations into funding decisions than a more narrowly focused review and is especially useful for research with a direct impact on society.

3

PREVENTING THE MISUSE OF
RESEARCH AND TECHNOLOGY

"ANYONE HERE WHO DIDN'T TAKE THE
HIPPOCRATIC OATH?"

Predicting the future course and consequences of research can be difficult or impossible. The development of nuclear weapons grew directly from fundamental research into the properties of subatomic particles. The technology known as genetic engineering emerged from research into antibiotic and virus resistance. History has demonstrated that the unfettered pursuit of new knowledge has many benefits that cannot be anticipated at the time the research was done. Nevertheless, research in some areas poses risks, and these risks need to be anticipated and minimized to the extent possible in the planning, performance, and dissemination of research.

The difficulty of predicting the future course and applications of research does not absolve researchers of the responsibility for participating in venues to explore these issues. **Researchers need to participate in discussions about the possible consequences of their work, including harmful consequences, in planning research projects.** As the ones who design and carry out research, researchers can provide information on the nature and purpose of research that is not available in any other way. Society funds research with the expectation that new knowledge will deliver benefits to health, the environment, and overall well-being. It expects researchers to do what they can, within their roles as researchers, to see that the promise of research is realized.

Development of Guidelines

Researchers are responsible for participating in the creation of institutions and practices to address the possible risks of existing and emerging technologies. The Asilomar Conference on Recombinant DNA is an example of researchers exercising this responsibility. One of the most notable advances in the life sciences in the past half century was the development during the 1970s of techniques for combining genetic material from different organisms. As this work was progressing, some leaders in the field warned of possible hazards from this line of experimentation. One concern, for example, was that bacteria incorporating genes from tumor-causing viruses might be dangerous to researchers working with these materials or to the broader public if released into the environment.

In response to these concerns, biologists held a four-day conference at Asilomar, California, in February 1975. The conference was able to generate substantial consensus on how to proceed to ensure the safety of various types of experiments, the experiments that should be deferred until more was known, and the priorities for steps to be taken by individual scientists, institutions, and national governing bodies to ensure safety in the future. Over the next several years, in cooperation with funding agencies and research institutions, the life sciences' research community took the lead in developing safety protocols for handling and sharing possibly dangerous materials, methodologies for assessing risk, and other practices and institutions that have proven to be highly effective and robust in protecting researchers and the public.

The Asilomar conference is an excellent example of action that benefits both the advance of knowledge and the public interest. Some research should not be performed because its limited expected rewards are not commensurate with its high risks. However, the development of guidelines to prevent or restrict research that can potentially harm society is still controversial within the scientific community. New challenges that cannot be predicted will continue to emerge with new advances. This is illustrated by the debate over whether research using new technologies that allow genetic editing of human embryos should be restricted or governed by new guidelines (Cyranoski and Reardon 2015).

Dual-Use Technologies

The term *dual use* traditionally described research-based technologies that have both peaceful and military uses. During the twentieth century, advances in nuclear physics, biology, and chemistry enabled the development of weapons capable of inflicting casualties on a massive scale. Through conventions such as the Nuclear Non-Proliferation Treaty

(NPT), the Biological and Toxin Weapons Convention (BTWC), and the Chemical Weapons Convention (CWC), the international community has sought to eliminate or constrain the use and spread of such weapons of mass destruction.[1] While the BTWC and CWC ban development activities aimed at producing weapons, these conventions do not directly restrict research.

In recent years, particular concern has focused on the potential for research in the life sciences and biotechnology to be used for harmful purposes. Knowledge, tools, and techniques developed in the life sciences have the potential to be misused for terrorism or to create new biological weapons (see figure 3-1). Also, in contrast to the development and manufacture of nuclear weapons, which requires an extensive human, physical, and technological infrastructure, some bioterrorism agents can be created by small groups or even individuals with the right training and access to certain facilities and materials. An example of such misuse is Aum Shinrikyo, the group that carried out a sarin gas attack on the Tokyo subway system in 1995 that killed thirteen and injured many more and that also experimented with bioterrorism agents. The 2001 anthrax mailings that killed five in the United States is another example of misuse. A U.S. biodefense researcher with access to advanced facilities was identified by the U.S. Federal Bureau of Investigation as the perpetrator after he had committed suicide, but some questions and uncertainties about the case and the FBI's conclusions remain (NRC 2011a).

---

1   These are formally known as the Treaty on the Non-Proliferation of Nuclear Weapons (NPT) (www.un.org/disarmament/WMD/Nuclear/NPT .shtml); the Convention on the Prohibition of the Development, Production and Stockpiling of Bacteriological (Biological) and Toxin Weapons and on their Destruction (www.unog.ch/bwc); and the Convention on the Prohibition of the Development, Production, Stockpiling and Use of Chemical Weapons and on their Destruction (www.opcw.org/chemical -weapons-convention/).

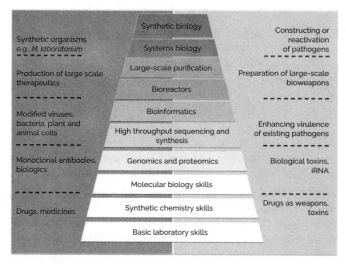

Figure 3-1. A hierarchy of increasingly advanced life sciences techniques (central pyramid) can be either used for the public good (left column) or misused to harm (right column).
Source: Flower 2011; NRC 2011c.

The *Statement on Biosecurity* from the IAP (2005) states that "scientists have an obligation to do no harm" and "should be aware of, disseminate information about, and teach national and international laws and regulations, as well as policies and principles, aimed at preventing the misuse of biological research." It also states that "scientists who become aware of activities that violate the Biological and Toxin Weapons Convention or international customary law should raise their concerns with appropriate people, authorities, and agencies." Governments, scientific associations, and research organizations around the world have created institutions, legal frameworks, and codes of conduct to prevent misuse of research results while ensuring continued scientific progress. For example, the Royal Netherlands Academy of Arts and Sciences (KNAW 2008) developed a

Code of Conduct for Biosecurity that describes the responsibilities of individual researchers, research institutions, journal editors, and others. A comparable document is being prepared at the Indonesian Academy of Sciences with assistance from the Netherlands. The Science Council of Japan endorsed including dual-use issues in its code of conduct for scientists in January 2013 (SCJ 2013), and the U.S. government has developed a policy for the oversight of dual-use research of concern in the life sciences in response to the 2012 controversy over studies of the transmissibility of the H5N1 influenza virus (USG 2012, 2014).

Discussion of issues associated with research on dual-use technologies need to be incorporated in the early stages of a researcher's training (NRC 2011b), and educational resources focused on fostering responsible behavior are available (see box 3-1). One example is an effort carried out by the National Research Council to develop effective methods for teaching about dual use technologies in developing countries. A series of workshops have been held that involve interaction with experts as well as the use of case studies and role playing (NRC 2013).

Box 3-1. Focus: Examples of Educational Resources

Federation of American Scientists (USA), Case Studies in Dual Use Biological Research
> Includes background materials, interviews with researchers and primary scientific papers, and discussion questions (in English, French, and Chinese)
> (Available at: http://www.fas.org/biosecurity/education/dualuse/index.html)

International Union of Pure and Applied Chemistry, Multiple Uses of Chemicals

Resource materials to help teachers and students understand the multiple uses of chemicals, learn about the Chemical Weapons Convention, and develop codes of conduct to prevent harmful uses (slides in English, four background papers available in Arabic, Chinese, English, French, Russian, and Spanish)
(Available at: http://multiple.kcvs.ca/)

University of Bradford (United Kingdom), National Defense Medical College (Japan), Landau Network Centro Volta (Italy), Education Module Resources (EMR)

Twenty-one lectures, with accompanying notes for the lecturer and direct links to references and videos; can be tailored for individual and more limited uses (in English, Japanese, Russian, Spanish, French, Urdu, Romanian-Moldovan, Georgian, and Polish)
(Available at: http://www.brad.ac.uk/bioethics /educationalmoduleresource/)

University of Bradford (United Kingdom), National Series.

Designed to help facilitate the immediate introduction of dual use biosecurity education into the higher education and professional science sectors of partner countries, in part by developing country-specific educational material and teaching guidelines; as of late 2013, a five-lecture set with accompanying teaching materials was available for twenty countries
(Available at: http://www.brad.ac.uk/bioethics/national series/)

> **U.S. National Academy of Sciences/National Research Council (United States), Developing Capacities for Teaching Responsible Science in the MENA Region: Refashioning Scientific Dialogue.**
>
> Reports on education institutes designed to create networks of faculty able to teach about dual use issues in a broader framework of responsible conduct and using the most advanced teaching methods focused on active learning
>
> (Available at: http://www.nap.edu/catalog.php?record_id=18356)

Preventing the misuse of life sciences research will continue to challenge researchers and the broader research enterprise. The increasing convergence of chemistry and life sciences, as well as the involvement of the physical sciences and engineering in fields such as synthetic biology, mean that these challenges extend well beyond biology. Tensions exist, for example, between traditions of open publication and sharing of scientific results and proposals for restrictions or government oversight in response to security concerns (see box 3-2).

> **Box 3-2. Focus: Gain-of-Function Experiments**
>
> In 2012, *Science* and *Nature* published research on how a deadly strain of influenza virus could be modified so that it could be transmitted by air to ferrets (Herfst et al. 2012; Imai et al. 2012). These so-called gain-of-function (GOF) experiments involve the modification of infectious agents to enhance their pathogenicity, transmissibility, or host range. Observers

raised concerns that the publication of the papers could enable malevolent misuse. Others argued that advancing such knowledge would be essential for understanding and combating future flu pandemics. Although these particular papers were eventually published, the broader tension remained (Imperiale and Casadevall 2014).

During the summer of 2014, two incidents involving inappropriately handled infectious agents at the U.S. Centers for Disease Control and Prevention served to intensify this debate. The White House Office of Science and Technology Policy (OSTP) and the U.S. Department of Health and Human Services announced the launch of a detailed review of GOF studies (OSTP 2014) and imposed a pause in new funding for this type of research while a new policy was developed.

The ongoing debate over where and how to strike the balance between the benefits and risks in GOF research is likely to continue within the microbiology community, including researchers, research institutions, research sponsors, and journals. For example, how can the risks of research be more thoroughly examined at the proposal stage as opposed to the publication stage? The questions and issues extend beyond GOF studies to other research on deadly agents. For example, in 2013 U.S. scientists discovered a new strain of botulinum toxin related to previous strains that have been used in biological weapons and are covered by both the BTWC and CWC. Because of the potential security risks and because no treatment for the new strain was available, the *Journal of Infectious Diseases* agreed to publish reports of the discovery but withheld the genetic sequence while an antidote was being sought (Barash and Arnon 2013). An accompanying commentary discussed the ethical issues associated with the decision to withhold some of the data (Relman 2013), an approach that has been recommended for controversial dual use cases.

4

CARRYING OUT RESEARCH

"IT WAS MORE OF A 'TRIPLE-BLIND' TEST. THE PATIENTS DIDN'T KNOW WHICH ONES WERE GETTING THE REAL DRUG, THE DOCTORS DIDN'T KNOW, AND, I'M AFRAID, NOBODY KNEW."

Research can take an endless variety of forms. It can involve collecting data with large instruments, theoretical inquiry, conducting surveys of human populations, imaginative brainstorming, creating numerical models, close study of texts or human artifacts, and many other activities. Finding a common thread across such diverse activities can be difficult. But an important observation is that research is both an individual and a social undertaking. Researchers must convince not only themselves but also their colleagues that their conclusions are reliable and valid. To do so, they must present the evidence and the reasoning behind their conclusions to others so that their contributions become part of the accepted body of communal knowledge.

Data Collection and Analysis

The forms of data used in research are as varied as research itself. Data can include text, numerical information, images, or video and audio recordings. They can be derived from observations of phenomena or objects, from experiments, or from computer simulations. They can be generated specifically for research or gathered from other sources. They can be continuous, discrete, or summarized in metadata. **Given the primacy of data in research, researchers have an obligation to keep clear, accurate, and secure records of their research data and of corresponding primary material so that their work can be verified or replicated by others**. Data are stored in many different forms, from notes on paper to recordings on magnetic tape to digital media. Research groups and institutions often have policies and procedures for protecting the integrity of data. Fields of research may also have unwritten procedures and professional standards that are passed from one generation of researchers to the next through example and mentoring. Many groups within the research enterprise are involved in the collection, processing, storage, and dissemination of data, but researchers bear the ultimate responsibility for ensuring the integrity of research data they generated or whose generation they supervised (see box 4-1).

Data typically undergo successive rounds of analysis and processing over the course of a research project. When data are processed, techniques used in that processing, including the computer code used to do that processing, become crucial to a full understanding of the data. It may even be necessary to provide others, including peer reviewers, with a thorough representation or copy of the devices used to do the processing for others to understand and replicate how data were processed. Indeed, many journals stipulate that referees or editors can require this level of access as part of the decision-making process.

Data need to be retained for sufficient periods to verify research results. For example, research sponsors may require institutions to keep data for a certain period of time as a condition of a contract or grant. Rights and responsibilities related to research data vary considerably according to the laws of individual countries.

Rapid and pervasive advances in information technologies are changing both the pace of research and how it is done. Digital data have become critically important in many fields, with significant implications for the research enterprise.

Data Sharing and Openness in Research

Research is built on openness. By making their conclusions and the evidence on which those conclusions are based public, researchers allow others to verify and build on their results. **Researchers are expected to share their data with others, including, where feasible, the research materials and software that enable them to draw their conclusions**. Providing access to data, algorithms, and software is important in areas of research where results cannot be duplicated, such as research on natural phenomena and simulations involving random processes. In the life sciences, providing access to research materials covers such materials as reagents, plant germplasm, cell lines, model organisms, and the means to rederive these materials (NRC 2003). Researchers who fail to meet these expectations place their reputations at risk (see box 4-1).

It may not always be possible to share all the data generated in research. Research results that contribute to a commercial product may need to be kept private while the product is being developed or until a patent can be prepared (although much research conducted by the private sector could be made publicly available without the loss of commercial advantage). Some types of personally identifiable information collected in the course of research are sensitive and need to remain private to prevent harmful uses of that information. Confidentiality is mandatory for individuals' medical data. Some data cannot be openly published because of national security concerns. In cases where researchers cannot release the data justifying their conclusions, they should be prepared to explain why data are not being released, and journals may require the provision of such explanations as a condition of publication.

If data cannot be released publicly, researchers may need to seek other ways of submitting their results to the judgment of peers. In industry or in military research, for example, other researchers and research managers may be able to review data that cannot be publicly released.

In some fields, researchers may temporarily be given exclusive access to a dataset as an incentive for gathering those data. Such provisions typically are negotiated within a discipline or interdisciplinary field to balance the benefits of openness with the harms caused by limiting the spread of information. This period of exclusive access should be as short as possible to enable subsequent verification and extension of results.

A primary benefit of making data, code, and other information underlying results accessible is that other researchers can try to duplicate the work. Data availability allows honest errors to be uncovered more quickly. In addition, the expectation or requirement that data will be shared can serve as a deterrent to fabrication or falsification in some fields. Finally, transparency in the area of data and analysis can help to discourage irresponsible practices that fall short of fabrication or falsification.

Reproducibility is one of the cornerstones of the scientific method. The problem of irreproducibility of research results is attracting increased attention in the scientific and general press (*Economist* 2013; Prinz et al. 2011). A 2013 article reported that most landmark preclinical cancer studies are not replicable and suggests several areas of focus for improving reproducibility, such as repetition of basic experiments, presentation of all results, and care to ensure that statistical tests are appropriate (Begley 2013). Just because a result is not reproducible does not mean that irresponsible behavior has occurred. Pinpointing the causes of irreproducibility in research and developing ways to foster reproducible research are long-term tasks for all participants in the global research enterprise, including research institutions, journals, and funding agencies (*Nature* 2013).

Researchers are taking on problems that they have not been able to address before, including difficult societal problems, and are making data available to other researchers and to the public. Individuals and research groups generate large databases that unrelated investigators may find relevant to their research. A key organization working to improve the quality and accessibility of data at a global level is the Committee on Data for Science and Technology of the International Council for Science (CODATA 2013). Global data efforts are being undertaken at the disciplinary level as well. For example, the Global Earth Observation System of Systems (GEOSS) being built by a coalition of governments and international organizations is designed to improve access to earth observations data to help protect against natural disasters, respond to climate change, and serve other needs (2013).

New technologies also are enabling researchers to make use of data in ways not possible before. For example, analysis of immense databases can reveal unexpected relationships that provide insight into causal relationships. Researchers now can post large databases online, widely disseminate research results online, and participate in widely available public forums outside the traditional peer-reviewed literature. Research fields may need to develop new methods of reviewing results and arriving at consensus to deal with such changes.

A new approach to research enabled by information technologies extends the openness of small research groups to much broader communities. By posting a problem in an open electronic forum and inviting contributions from anyone who wants to work on the problem, many different perspectives can be brought to bear on a problem. This approach raises several interesting questions, such as how and whether individual researchers should gain credit for their contributions to solving a problem and how the resulting research should be published.

Making data, code, and other information available in a way that they can be used by others can be time consuming and resource intensive (Molloy 2011). Researchers who are not required to share data by funder, journal, or institutional mandates (see the following) may be reluctant to provide access to unique data. Lowering barriers to data sharing is an important task for all stakeholders in the global research enterprise.

Active curation and preservation of data often are necessary to ensure that the full value of data is realized, even the unanticipated value. Some fields of research have well-developed procedures for determining how best to preserve data and institutions to carry out preservation, but others do not. Funding agencies such as the National Institutes of Health in the United States and the Wellcome Trust in the United Kingdom have mandated the public deposition of all data generated during the funded research (Van Noorden 2014a). In addition, journals such as PLoS are promoting open access to the data underlying papers they publish. Given that curation and preservation involve long-term costs, discussions are under way at the global and national levels among research institutions, sponsors, journals, and others over how to accomplish preservation and how to pay for it.

Differences in research methods also can create complications in interdisciplinary collaborations. Agreement on accepted methods before an interdisciplinary project starts can help prevent later difficulties.

Fostering Broad Participation in Research

Openness in research implies a willingness to consider multiple contributions to research. A successful research system embraces and encourages the contributions of groups that are underrepresented in particular fields, including women, underrepresented ethnicities, and people with disabilities. Multiple perspectives can speed and broaden research, and the members of all groups can make vital contributions to human knowledge. Research communities in individual countries and internationally have been emphasizing the need to increase the participation of women and other underrepresented groups in science and technology fields (IAC 2006; NRC 2011). Efforts are motivated both by the moral imperative that occupations within society should be open to all, as well as by the practical understanding that a larger potential pool of skilled scientists, engineers, and medical professionals is needed to strengthen discovery, innovation, and health around the world (box 4-1). As specific examples, women who have had an enormous impact historically and whose contributions underlie modern science include Marie Curie (physics and chemistry), Lise Meitner (physics), Barbara McClintock (biology), Dorothy Hodgkin (biochemistry), Rosalind Franklin (biophysics), and Anne McLaren (developmental biology).

Indigenous and traditional knowledge systems should be respected for their potential contributions to human understanding and well-being. For example, local populations

---

Box 4-1. Discussion Scenario: Inclusion

You are a graduate student at a large research university. You learn from the principal investigator of your research group that a new graduate student will be joining the group, and that this student is a member of an ethnic group in your country that is underrepresented in research. Moreover, the new student has less training than other members of the group, and the PI asks the group for their help in getting him up to speed. In a later conversation where the principal investigator is not present, several members of the group disparage the new student's ethnicity and express doubts about his ability to contribute.

How would you react? Would you confront those who are making disparaging remarks? Would you speak with the principal investigator? How would you behave toward the student when he joins the group?

---

have irreplaceable information about the local biota and environment, and many pressing social problems cannot be solved without incorporating information from businesses, farmers, community partnerships, and other local sources (Nyong et al. 2007). This indigenous knowledge should be subjected to the same rigorous testing as other scientific hypotheses, but researchers cannot assume that only one pathway leads to knowledge.

Sometimes research results lead in unexpected directions and overturn existing worldviews. Researchers should welcome new results despite their potential to affect existing research programs. At the same time, novelty is greatly valued in research, creating incentives to trumpet new results as novel without justification. Successfully balancing the desire for novelty against the cumulative weight of past research is one measure of a good researcher.

The methods and conclusions of research apply to the entire intellectual world, and scientists and scholars from throughout the world can and should participate in this universal activity. Those who argue that the values of science are the product of a particular cultural, religious, or geographic perspective deny this universality.

Actions that Damage the Research Enterprise

Irresponsible practices in the conduct of research can take many forms and can occur throughout the research process. Among the most egregious are those that violate the trust underlying research by introducing fraudulent results into science or scholarship or by stealing ideas. These acts include fabrication, which is making up results and recording them as if they were real; falsification, which is manipulating research processes or changing or omitting data; and plagiarism, which is appropriating another person's material (including ideas, research results, or words) without giving proper credit. **Researchers have an obligation to themselves, their colleagues, and society to avoid the egregious transgressions of scientific values, including falsification, fabrication, and plagiarism and other forms of irresponsible conduct that can undermine the research enterprise.**

Irresponsible research behavior can include what may seem like minor transgressions. For example, researchers might discard outlying data from experimental results under the assumption that something must have gone wrong with the experiment to produce the outliers. They may alter images in such a way as to emphasize some aspects of the image and deemphasize others (box 4-2). However, such actions violate researchers' fundamental obligation to produce reliable and objective results. Figures that do not represent the actual data obtained in the lab, for example, can lead other researchers astray.

Box 4-2. Focus: Avoiding Inappropriate Manipulation of Digital Images

The use of digital images to record the results of experiments and the wide availability of software to edit or alter these images have raised new issues and challenges for researchers. In many fields of research, images are a major form of experimental data. Purposely manipulating or altering images so that they do not represent research results constitutes fabrication or falsification. The U.S. Office of Research Integrity (ORI) reported that more than two-thirds of the research misconduct investigations opened during 2007–8 concerned questions about images (Gilbert 2009). In addition to image manipulation aimed at misrepresenting results, researchers may make inappropriate changes that are inadvertent or are motivated by a desire to render an image more aesthetically pleasing, with real or perceived pressure coming from journals (*Nature Cell Biology* 2006).

In response to these developments, a number of journals have clarified their policies on images and have begun checking a percentage of the images accompanying provisionally accepted papers for inappropriate manipulation. Nature Publishing Group (NPG) maintains a Web page with the policies it has established for its journals (NPG 2013), and material is available from other sources. For example, the University of Alabama at Birmingham (2008) has developed an online learning tool for digital imaging, which includes guidelines, videos illustrating the guidelines, a case study, and other materials.

Several general guiding principles can be drawn from this material:

- Maintain the original image, and consider images a form of data.
- Images should be minimally processed—the less the better.

- In cases where image manipulation is appropriate or unavoidable, keep track of all image-processing steps, and describe all these steps, along with the software used in the methods section of the paper.
- In general, cropping and simple adjustments applied to an entire image are acceptable.
- Manipulations that affect only one part of the image are problematic.
- Specific questionable practices that are not recommended include the use of Photoshop's clone stamp tool to "clean up" an image, the use of filters, and the use of "lossy" compression (such as the technique used to produce jpeg files, which involves the loss of some data).

Fabricated or falsified research can be extremely harmful to researchers and to society (box 4-3). It can result in the production of deficient products, inadequate instruments, or dangerous or ineffective medical therapies or drugs. Policy or legislation can be based on incorrect findings. The public's trust in science and scholarship can be damaged. The fabrication or falsification of results can end a researcher's

Box 4-3. Focus: The Stapel case

The case of Diederik Stapel demonstrates the extent to which fraud can damage colleagues, a field of research, and science in general. A social psychologist in the Netherlands, Stapel studied social processes such as how prior experiences, including exposure to a priming word or concept, could affect responses to a situation. Early in his career he began making up the data supporting his published papers, including papers that were published in prominent journals and attracted widespread media attention. His falsification of data was discovered when three graduate students began to notice anomalies in the data and could not get satisfactory answers to their questions.

Three investigative committees set up to investigate Stapel's work at the three universities where he was employed concluded in their final report that he had fabricated data in at least 55 of his 137 papers and in the PhD theses of 10 students he supervised. In other papers, the data were missing, but statistical analyses showed evidence of fraud. The final report of the investigative committees pointed to the devastating effects that Stapel's fraud had on his former PhD students and postdoctoral fellows, who did not participate but "whose publications were suddenly becoming worthless." The report also faulted the field of social psychology and the institutions involved: "journals, reviewers, assessment committees and graduate schools must learn a methodological lesson: it appears that too much can go wrong in the discipline's critical functions" (Levelt Committee et al. 2012).

The Stapel case generated a shock wave not only through social psychology in the Netherlands and abroad but also in other parts of science. The *Flawed Science* report (Levelt Committee et al. 2012) functioned as a wake-up call for universities and research institutes to take a number of measures to guard against both fraud and sloppy scientific practices (see Drenth 2013).

career, discredit colleagues, and damage the entire research enterprise. The retraction of published papers in a research area tends to decrease subsequent funding for research in related areas (Azoulay et al. 2012).

Plagiarism may seem to be a less severe transgression than fabrication or falsification, and it does not have the effect of introducing fraudulent results into research.

However, because it is based on deceiving other researchers, it, too, weakens the foundation of trust on which research is based. Also, when plagiarism is accepted or tolerated it can sap the motivations of others and undermine incentives to be open, since reward systems in research are largely based on credit for publications (box 4-4). Electronic communications have made it both easier to plagiarize material and easier to detect plagiarism, but such means of communication have not changed the expectation that published material is not copied from elsewhere unless noted and properly referenced.

Publishing or duplicating exactly the same material in more than one place—a practice known as duplication or by the incorrect term "self-plagiarism"—is dishonest when publishers and readers expect published material to be original, and it squanders the resources of authors, reviewers, editors, and publishers (see also chapter 2). At the same time, ongoing discussion in the publishing and ethics communities involves whether duplication of some technical sections of a paper should be considered acceptable. In any event, proper reference to the original publication is mandatory. In addition, when researchers publish their work in journals outside their native language, they need to ensure that the descriptions as well as the work reported are original, unless noted, and that the translation is correct.

Sometimes there are gray areas where it is difficult to determine if a researcher has engaged in falsification, fabrication, or plagiarism. For example, a researcher may use methodologically unsound data processing, questionable analytical or statistical techniques, or inadequate control groups. A case study may not be representative of the phenomena it is purported to represent. Economic, ideological, or personal interests may skew the outcomes of research or peer review. Plagiarism may range from the wholesale theft of long passages of text to the careless or perhaps inadvertent use of another's ideas. All these cases

Box 4-4. Discussion Scenario:
Providing Credit in Research Projects

You are a postdoctoral researcher in the Division of Cancer Biology, and your advisor has invited you to collaborate in a research project. She chose you because the topic of the research matches your interests and because you are very effective at doing lab work. A few weeks later, you have become very involved in the project by designing experiments and compiling data, which involves extra hours in the lab, including weekends. About eight months later, you run some preliminary analysis of the data, and results look more promising than expected.

You schedule an appointment with your advisor to communicate your findings. Your advisor looks at the results, "This is great," she says. "Give me a couple of weeks to study the data more in detail and if everything looks fine we can start working on a draft manuscript for submission." You agree and spend the next few weeks working on other projects. A month later, without further discussion with you, your advisor presents the results at the monthly department meeting without mentioning your name.

Should your supervisor have mentioned you? What will you say to her when you next meet? How could this situation have been avoided? If your advisor had gone forward and submitted a manuscript for publication, reporting the work but did not list you as an author, what would you do?

represent irresponsible behavior, but it may be impossible to determine whether the researcher in question set out to deceive.

Standards and expectations can vary by discipline and change over time. Disciplines may have different mechanisms for attributing credit to individual researchers. The research and higher-education systems in some countries and some disciplines are more hierarchical than in others. Using significant blocks of text from one's own thesis or dissertation without quotation marks in subsequently published journal articles may not have been considered an irresponsible practice in the past in certain fields, whereas today it would if the thesis was formally published. A sense of fairness and proportion should be maintained when evaluating research behaviors. Many research fields need to accommodate a wide variety of approaches to make progress. Furthermore, researchers are human and can make mistakes. These mistakes need to be openly acknowledged and corrected so that other researchers do not subsequently act on the basis of incorrect information.

Beyond fabrication, falsification, and plagiarism are a host of irresponsible actions that may not involve the intent to deceive but nevertheless can damage the integrity of research results. Inadequately managing and storing data, withholding data from colleagues who want to replicate the findings, abuse of research subjects, lack of care in designing or undertaking experiments, misrepresentation of interests, inappropriate authorship, covering up irresponsible behavior, or reprisals against whistle-blowers who report irresponsible behavior that transgresses institutional or funder policies all constitute irresponsible practices in research. Making an indiscriminate or malicious allegation of irresponsible research practices is itself irresponsible and damaging and can stymie progress. Bringing such actions to light can help prevent or correct them. For example, discussing concerns in public, rather than in private communications, results in a sevenfold increase in retractions and corrections by journals (Van Noorden 2014b).

Finally, some irresponsible actions may not damage the research record but are inappropriate in any workplace. These include intimidating or harassing students or assistants, inadequate mentoring or counseling of students,

Box 4-5. Discussion Scenario: Data Sharing

As part of your postdoctoral research, you are conducting a meta-analysis that requires collecting data published by other researchers. In order to perform the study, you contact approximately 100 researchers in your field and request the raw data used in their original publications. After several months only about twenty authors have provided the requested data. The remainder either did not respond or replied that their data were incomplete or no longer accessible or that they were too busy.

What are the implications of this response for your field? What new approaches could be instituted to encourage data sharing? How would you ensure that all data generated from your research is properly maintained and can be shared?

misrepresentation of credentials, insensitivity to social or cultural norms, prejudice against members of particular groups or genders, misuse of funds, failure to disclose conflicts of interests, and other breaches of general social and moral principles. Some of these actions may be subject to criminal or civil prosecution and sanctions. In other cases, procedures similar to those used to respond to fraudulent actions in research may be used to investigate claims and respond appropriately. All workplaces should have procedures for dealing with these kinds of irresponsible actions.

5

THE RESEARCHER'S RESPONSIBILITIES TO SOCIETY

"As they say, 'Do the right thing.'"

In addition to their responsibilities to help prevent the misuse of research that were covered in chapter 3, researchers have additional responsibilities to society when they plan and perform research. Some of these responsibilities have been codified into laws and regulations, and research institutions and countries have created bodies to ensure that researchers adhere to these provisions. Other responsibilities are less well defined but nevertheless exert a powerful influence. Researchers who transgress these boundaries can do great harm to their careers and to research in general.

Some research can only be described as injurious to human welfare and cannot be justified, such as research focused on the development of weapons outlawed by international treaties. Research also is unacceptable if it conflicts with the basic human values of autonomy, freedom, dignity, nondiscrimination, and a lack of exploitation. Documents such as UNESCO's Universal Declaration on Bioethics and Human Rights (2005) and the related Bioethics Core Curriculum (2008) are useful resources in learning about how societal considerations bear on research in the life sciences.

Social and cultural considerations dictate whether some forms of research are acceptable, and these considerations can vary from one place and time to another (see box 5-1). In cases where research raises ethical issues, the research community should expect and welcome input from society while working with nonscientists to make that input as useful as possible. Such interactions will build trust between scientists and the public, including policymakers. Institutions may be needed to serve as brokers of information exchanges between researchers and policymakers so that policies can be informed by the best scientific advice.

Box 5-1. Focus: Research on Human Embryonic Stem Cells

Attitudes toward embryonic stem cell (ESC) research vary by country. Some countries (such as Austria, Ireland, and Italy) prohibit or heavily circumscribe it, while others (such as China, South Korea, and India) are less restrictive (Matthews 2007). Obviously, whether and to what extent ESC research can be accepted depend on social and cultural considerations that can vary from one place and time to another.

In the United States, for example, the Clinton administration supported ESC research following the 1998 breakthrough that enabled ESCs to be isolated and grown in cell culture. In 2001 the Bush administration issued an executive order that restricted federal government support for ESC research to already derived cell lines, meaning that any research on newly derived cell lines required support from private foundations or state governments. In 2009, the Obama administration lifted these restrictions. The shifting policies of the U.S. government reflect a clash in values. For adherents of some religious traditions, research use of ESCs that involves the destruction of human embryos is unethical. The Bush administration's policy reflected this view. For others, the promise of the medical advances that might result from ESC research should be weighed more heavily. Reflecting the latter view, President Obama (2009) stated that his administration will vigorously support scientists who pursue this research and will "aim for America to lead the world in the discoveries it one day may yield."

China presents a contrasting example in which there has been no clash in values over ESC research. China's population as a whole does not hold religious, ethical, or other beliefs that would be an obstacle to such research. The Chinese government considers ESC research to be a strategic, emerging technology and has established several national science and technology programs to support its development. In December 2003 the Ministry of Science and Technology and the Ministry of Health issued *Ethical Guiding Principles on Human Embryonic Stem Cell Research* to codify the ethical principles guiding China's ESC research (Ministry of Science and Technology and the Ministry of Health 2003).

Many complex issues characterize the relationship between research and society. This guide cannot cover all of them, so it provides a brief examination of several important topics—the protection of research subjects, bioprospecting and biodiversity research, privacy, and laboratory and environmental safety—to illustrate the scope of these issues.

Protecting the Subjects of Research

Regulations governing the performance of research cover human subjects, laboratory animals, laboratory safety, environmental protection, and other areas. Laws and regulations vary greatly by country, but some areas are more harmonized internationally.

Human subjects' protections constitute some of the most important legal and ethical rules that researchers must follow. The *Nuremberg Code* (1949) contains ten points defining ethical medical research, including the principle that "the voluntary consent of the human subject is absolutely essential." The World Medical Association's *Declaration of Helsinki*, first adopted in 1964 and revised several times since, and the Council for International Organizations of Medical Sciences' *International Guidelines for Biomedical Research Involving Human Subjects* are more extensive global codes of medical research ethics (WMA 2008; CIOMS 2002). The work of the Committee on Bioethics (DH-BIO) of the Council of Europe covering biomedical research and medical practice is also relevant, including the legally binding Oviedo Convention on Human Rights and Biomedicine and its protocols on issues such as genetic testing, as well as recommendations concerning xenotransplantation and research on biological materials (COE-DH-BIO 1997, 2003, 2006, 2008). Protections such as the requirement for informed consent of human subjects and the use of expert committees to review and approve research protocols have been extended to the social and behavioral sciences.

Special rules cover research involving prisoners, children, and other vulnerable groups.

An important area of research involving human subjects is the clinical testing of drugs and biologics (see also chapter 7). International collaboration in such clinical trials is growing rapidly. The U.S. Food and Drug Administration (FDA) reported in 2008 that 80 percent of approved applications contained data from outside the United States (DHHS-OIG 2010). For pharmaceutical companies and other entities financing clinical trials, conducting studies in developing countries delivers advantages such as lower cost and access to populations that have not previously been exposed to treatments. For developing-country participants, clinical trials may deliver access to care and treatment that would not be available otherwise.

FDA requires that research data supporting an application to market a new drug or biologic in the United States must be generated by trials that adhere to U.S. policies and procedures that protect human subjects. However, ensuring that ethical requirements are followed in some developing countries can be difficult (box 5-2). Further, the significance of concepts such as informed consent may be questioned in cultures where trust in medical professionals may be higher than in others or where declining to participate in a study would essentially mean forgoing medical treatment. International medical research also has seen egregious abuses in the past, such as U.S.-led experiments in Guatemala during the 1940s in which some patients were deliberately infected with syphilis and other sexually transmitted diseases without informed consent.

Today, efforts are underway to use international collaboration in clinical trials as a mechanism to build capacity in clinical medical research and prevent irresponsible conduct in developing country hosts. One example is the Institute of Human Virology Nigeria (IHVN 2004), which in partnership with the University of Maryland School of Medicine,

Box 5-2. Discussion Scenario:
Protection of Human Subjects

You are a health researcher or institutional official at a medical facility in a developing country. A multinational pharmaceutical company approaches you to see if your facility would be interested in participating in a multisite clinical trial of a new treatment for a childhood illness. At several developed country sites, the study would involve two trial arms, with one set of subjects receiving the new treatment and the other receiving an existing approved treatment. At one of the sites in a developing country, the company would like to conduct a three-armed study, with one group of subjects receiving the treatment being tested, the second receiving the approved treatment, and the third receiving a placebo. Because administration of the approved treatment is the standard of care at the developed country sites, using a placebo arm would be considered unethical at those sites. However, the approved treatment is unavailable in your country because of its expense, and use of the placebo arm at your site would allow the trial to be conducted more quickly.

What is your reaction to this plan? What if the company committed to donate a large supply of the existing approved treatment (or the new treatment if it is approved) to your facility and to other health clinics in your country?

government, and stakeholders seeks to provide high-quality health care that is accessible to all Nigerians. Efforts such as the Horizon 2020 Program of the European Commission provide funding to foster ethical conduct in biomedical research conducted in developing countries.

Researchers are also responsible for ensuring the humane treatment of animals used in experiments. In some countries the treatment of laboratory animals is highly regulated; in others, it is not. The *Guide for the Care and Use of Laboratory Animals* produced by the U.S. National Research Council is a widely used resource, even in countries with less-extensive regulations in this area (NRC 2011a).

Bioprospecting and Biodiversity Research

Over recent decades, the need to preserve and understand Earth's diverse plants and animals has become increasingly apparent (CBD 2014). Moreover, countries with rich biological resources have put increasing emphasis on ensuring that benefits generated through bioprospecting and related activities are appropriately acknowledged and shared. The discovery and commercialization of biological resources in pharmaceuticals and other industries has a long history. Since the traditional knowledge of indigenous people often enables the utilization of these resources, bioprospecting becomes exploitive when this knowledge is used without permission or appropriate benefits.

Policies on the collection, transfer, and utilization of biological resources differ among nations. Striking the appropriate balance among various interests, including the need for increased knowledge about these resources, the need to protect them, and the need to generate and appropriately share benefits, can be difficult. Researchers need to be aware of any national and international treaties that regulate the collection of biological material in relation to their work. Obtaining permission to legally collect such material is critical and should not be ignored (see box 5-3).

As an example, about 10 percent of the world's biodiversity exists in Colombia (Andrade-C et al. 2012; Fog 2011). The Colombian Constitution, promulgated in 1991, and Law 70 of 1993 contain provisions that protect the rights of ethnic communities to preserve their culture and territory. Furthermore, Colombia established regulations aimed at ensuring that the benefits accruing to the use of biological

Box 5-3. Discussion Scenario: Performing Fieldwork in a Protected Area

You are a biologist working on a PhD dissertation about marine protected areas and local communities. As part of your work, you travel to a national park in another country to collect samples of an endangered fish species and contact communities in the surrounding area. The permitting process has taken several months. Finally, you travel to the field after your paperwork is in order. During one of the scuba diving sessions, you spot a rare fish species that a colleague from your university would be interested to sample. However, that particular species is not included in your approved permit.

What would you do in this situation? Would it be acceptable to collect the sample and inform national park authorities later? What are the broader implications of illegal collection of biological material?

resources are shared according to the principles of equity and justice. Overall, these laws and regulations are seen as having benefited the affected communities and broader Colombian society.

However, Colombian researchers reported that the actual operation of these regulations made biodiversity research in Colombia nearly impossible (Andrade-C 2012). A range of permits and other documentation were required before researchers were allowed to do fieldwork, collect specimens, or even transport or import existing specimens collected many years before the law went into effect (Fernandez 2011). Among other permissions, researchers were required to obtain consent from all indigenous and African descendent communities in the sites to be sampled. According to butterfly researcher Miguel Gonzalo Andrade, a research program on Colombian butterflies would have required permits from more than sixty different communities (Andrade-C 2013). The entire process could take years and thousands of U.S. dollars for each project, and Colombian scientists bypassing these requirements were subject to criminal legal sanctions (Government of Colombia 2009).

Over the course of several years, Colombian scientists and research institutions worked to convince policymakers to simplify the regulations governing biodiversity research (Ministry of Environment and Sustainable Development 2013). In June 2013 the Colombian government issued two decrees (Decree 1375 and Decree 1376) that eased permitting requirements (Government of Colombia 2013a, 2013b). Going forward, higher-education institutions can apply for a "framework permit," valid for ten years, that covers individual scientists and projects related to a research program (El Espectador 2013). These changes have greatly facilitated basic, noncommercially motivated biodiversity research in Colombia.

Privacy

As the amount of data generated and stored about our daily lives becomes ever more voluminous and freely available, issues of privacy are growing in importance. For instance, the analysis of social media data by corporate marketing departments as well as academic social scientists raises concerns about the privacy of those whose online behavior is tracked. Genomics and medical research also raises many privacy concerns. As an example, the Icelandic biotechnology company deCODE used a large database of genomic information on Icelandic volunteers to identify the genetic risk factors associated with various diseases. In 2003 the country's Supreme Court blocked the company's attempt to work with the Icelandic government to develop a database of genetic data from all of Iceland's approximately three hundred thousand citizens (Gertz 2004).

Researchers working with personally identifiable information need to be aware of the regulations and laws that apply to the collection and use of such data. These regulations vary by country or region and are being reexamined and revised in some places. For example, updates to the European Union's rules protecting the right to not have one's personal data collected and used are currently being developed (EC 2015). The proposed General Data Protection Regulation would harmonize data protection regimes throughout Europe, specify the responsibilities of companies that hold personal data in the case of breaches, and address the implications of globalization, as well as social media, cloud computing and other new technologies.

Laboratory and Environmental Safety

In fields of research that involve the use of hazardous materials and equipment that could be dangerous to those performing research or others in the laboratory environment, the relevant laws, regulations, and protocols need to be understood and followed. In disciplines such as chemistry, creating and maintaining a culture that promotes laboratory safety is a shared responsibility of researchers, laboratory managers, and institutional officials (NRC 2011b). The specific regulations and safety procedures might vary by country and field. For example, implementation of laboratory safety and security protocols can be challenging in developing countries, since they may not be considered a priority, and regulatory frameworks may lack enforcement mechanisms (NRC 2010).

As a general recommendation, laboratory and safety procedures should be reviewed by researchers at least once a year (NAS-NAE-IOM 2009). Many research institutions require that employees attend training courses about the use of their research facilities, a common practice in most molecular biology labs. Moreover, other fields such as experimental physics utilize research facilities (e.g., particular accelerators) that require mandatory safety training to be completed before users can enter the facility.

Some areas of research carry risks of harmful impacts outside the laboratory, and risks need to be carefully evaluated before experiments can proceed. One example of such a research area is geoengineering—the deliberate alteration of the earth's atmosphere and other parts of the climate system to moderate the warming effects of greenhouse gases. Interest in geoengineering is increasing at a time when the prospects for greenhouse gas abatement efforts are uncertain at the global level and concerns about potentially catastrophic effects of climate change are growing. Large-scale efforts aimed at managing solar radiation or removing carbon dioxide from the atmosphere would require significant international public discussion and development of policy, in addition to considerable advances in knowledge of climate science and possible climate engineering approaches. Reports by the Royal Society (UK) and the U.S. National Academies explore these challenges and make recommendations on how they might be addressed (The Royal Society 2009; NRC 2015).

Other areas of research that have spurred debates about environmental safety include genetically modified crops and nanotechnology. For example, China identified nanotechnology as a priority research area in 2001, and since then it has launched a number of specialized institutes and provided increased funding. China has put particular emphasis on nanomaterials and nanodevices research, areas where it has emerged as a world leader, and it is quickly developing capability in such areas as nanobiology, nanomedicine, nanocharacterization, nanotoxicology, and nanobiosafety. This strategy has produced impressive results, with the number of nano-related papers written by Chinese researchers increasing at an annual rate of 23.8 percent between 2001 and 2010 to a total of more than twenty thousand, the highest growth rate in the world (Klochikhin and Shapira 2012). As a result, China ranked first in the world in the number of nanomaterial basic research papers appearing in the Science Citation Index (Arora et al. 2012; Chen et al. 2010; Qiu 2012) and has attained a significant number of patents.

Several Chinese institutions conduct nanosafety assessment research, including two Chinese Academy of Sciences (CAS) institutes: the Key Laboratory for Biomedical Effects of Nanomaterials and Nanosafety, and the Suzhou Institute of Nanotech and Nanobionics. Nanosafety research in China has made significant progress on such questions as the ability of nanoparticles to pass through biological barriers, the biotoxicological effects of nanoparticles inside the body, and the methodology of nanotoxicology.

Issues of scientific responsibility in nanotechnology also are receiving increased attention. For example, the Scientific Morality Construction Committee in CAS's Academic Division initiated a series of studies in 2011, including one titled "Ethical Issues in Nanotechnology and the Social Responsibility of Scientists." The project is exploring potential ethical and social problems of nanotechnology, governance mechanisms, and the responsibility of the nanotechnology R&D community. The end result of the activity will be "A Proposal for Responsible Conduct of Nanotechnology Research and Development." CAS also is working to build dialogue among researchers in the natural sciences, social sciences, and humanities around nanotechnology.

Safety and security procedures are essential tools for research facilities to be used effectively, and efforts to improve and implement these practices should be viewed as enablers rather than inhibitors of research.

6

PREVENTING AND ADDRESSING IRRESPONSIBLE PRACTICES

"MY RESEARCH COVERS TWO FIELDS: THE BEHAVIOR OF MATTER UNDER HIGH PRESSURE, AND THE BEHAVIOR OF SCIENTISTS UNDER HIGH PRESSURE."

Approaches to addressing irresponsible research practices generally fall into two categories: promoting research integrity and the adoption of good practices through education and training and handling irresponsible behavior through procedures to investigate alleged transgressions and impose penalties when allegations are confirmed (ICB 2010). Participants in the global research enterprise are coming to understand that these approaches are complementary. Preventing irresponsible behavior is preferable to corrective action after the fact. At the same time, the existence of robust and effective corrective procedures can serve as a deterrent to misbehavior, in addition to correcting the research record and identifying deficiencies.

Individual researchers, research institutions, research funders, journals, professional societies, and science academies all have important roles to play in ensuring the integrity of research. Most of these are covered in this chapter, but see chapter 8 to better understand the role of scientific journals

A number of codes and guidelines have emerged in recent years from a variety of countries and organizations around the world that spell out the responsibilities of these individuals and groups for ensuring integrity. These codes and guidelines differ in their details depending on cultures and specific legal and institutional frameworks. Nonetheless, they agree on many key issues.

The Incidence of Irresponsible Conduct

No one knows the exact prevalence of various kinds of irresponsible research practices. Available information from the countries that report statistics on investigations and surveys indicate that the vast majority of researchers are upright and honest. Still, while the most serious forms of irresponsible conduct, such as fabrication or falsification of data, are unusual, they occur frequently enough all over the world to merit focused attention from the global research enterprise. A meta-analysis of surveys found that about 2 percent of scientists admitted to having fabricated, falsified, or modified data or results at least once, and up to one-third admitted to engaging in other questionable research practices (Fanelli 2009). This meta-analysis also found that 14 percent of scientists had observed falsification by colleagues, and 72 percent had encountered other questionable research practices less serious than fabrication or falsification.

Trends in the prevalence of irresponsible research conduct also are uncertain. Retractions of published scientific papers, both in terms of the numbers and causes, have attracted increasing attention (Steen et al. 2013; Van Noordeen 2011; Fang et al. 2012). High-profile cases of scientific fraud continue to occur in countries throughout the world (Retraction Watch 2015).

Shifts in the research enterprise such as expansion of the research system, an accelerated pace of research activities, increased pressure to publish in prestigious journals, and heightened competition for funding can have the effect of weakening the transmission of codes of conduct from one generation to the next and mitigating against an environment that fosters responsible conduct. As pointed out in chapter 4, irresponsible research behavior can cause grave damage to individual researchers, to coworkers, to a scientific discipline, to science in general, and to society. Given these severe harms, the need to prevent and respond to irresponsible research practices is strong.

Responsibilities of Individual Researchers

Many of the responsibilities of individual researchers identified in various codes and guidelines, such as effective mentoring, upholding the truth in performing and reporting their own work, appropriately crediting others, and observing environmental and safety guidelines, are covered in the rules and guidelines of their own institutions and in other laws and regulations that bear on the responsible conduct of research.

In the ongoing dialogue among researchers on ethical issues (NHMRC 2007; ESF-ALLEA 2011), there has been discussion of whether scientists should develop an oath similar to medicine's Hippocratic Oath (Cressey 2007). Such an oath could perhaps be sworn by PhDs at the outset of their careers. However, experts who have seriously considered this suggestion have found it difficult to identify universal language that includes all scientists and situations (AAAS 2001).

Individual researchers play a variety of roles throughout their careers, with responsibilities generally increasing over time as they progress from student, to postdoctoral fellow, to junior researcher, to PI. At all times, researchers are responsible for inquiring about anomalous behavior and calling attention to possible irresponsible behavior given cause to do so. More senior researchers have the added responsibilities of mentoring junior colleagues in good practices and of observing standards of fairness and impartiality when participating in activities such as peer review. Researchers have a responsibility to check the work of coauthors in collaborative projects for possible anomalies and mistakes. Over the course of their careers, some researchers may play other roles that carry additional responsibilities, such as exercising administrative oversight, participating in the activities of societies and associations in their fields, being elected to an academy, or contributing to the allocation of resources for research as part of a funding organization.

Almost every researcher eventually will encounter instances of irresponsible conduct in research. Dealing with these situations can be very difficult. But how researchers respond to irresponsible conduct can have a major influence on their careers and on the strength of the research enterprise.

It can be very difficult to raise concerns about the actions of another researcher, especially when that person is in a position of authority. But researchers cannot uphold the fundamental values of research while ignoring irresponsible research practices. **Researchers have a responsibility both to maintain high standards of responsible conduct and to take appropriate actions when they witness or suspect irresponsible conduct.**

Many concerns can be addressed by talking with someone else within a research group, perhaps someone who has been designated as a point of contact on research practices. However, people who have concerns about the actions of another researcher need to have more than one way to make those concerns known.

Readers of this guide may wish to know more about the specifics of what they should do and whom they should contact to report irresponsible conduct such as fabrication, falsification, or plagiarism. If the allegation relates to published or submitted work, it may be appropriate to raise concerns with the journal. Many journals, institutions, sponsors, and government organizations that address allegations provide information about points of contact and procedures for raising questions. Researchers should familiarize themselves with those that apply in their situation. However, as discussed shortly, institutional and government approaches to irresponsible research behavior vary around the world, so it is impossible to provide specific guidance that will fit every case. Some institutions and countries do not have clear policies or designated points of contact, and no international agency exists to investigate allegations. Researchers should do what they can to raise awareness in their institutions and national research enterprises about the importance of responsible conduct and the need for clear procedures to address allegations of wrongdoing.

Responsibilities of Research Institutions

Research institutions also have responsibilities to prevent and address irresponsible research practices, and the policies they implement have a direct impact on researchers. All researchers, including those involved in collaborations with colleagues at an institution, should be familiar with these policies.

The first and most important aspect of attention to irresponsible conduct is its prevention. Research institutions need to ensure that all researchers, research staff, and students receive both formal and informal training in responsible research practices. All researchers need opportunities to learn about the values and principles on which good research is based. Responsible conduct should be an element of all courses and research experiences so that it is seen as fundamental to the research enterprise. Institutions are also using new technologies, such as software that can help detect plagiarism and image manipulation, to check dissertations and other work.

In addition, more institutions around the world, such as national academies for young researchers, are launching stand-alone programs in responsible conduct of research (RCR) education, mainly aimed at graduate students. Curricula, cases, and other teaching materials are proliferating, although much of this growth is still limited to the developed world and to the United States in particular. Research is being undertaken to determine which approaches to RCR education are most effective, and this work has uncovered some useful insights. For example, programs that are relatively more successful tend to rely on stand-alone seminars rather than embedding education in other coursework and use interactive, case-based teaching methods (Antes et al 2009). However, the evaluation of RCR education presents significant challenges, as does the evaluation of education programs in general. Much remains to be learned.

In China, researchers, research institutions, and funding agencies have been grappling with issues of research integrity in recent years, as the size of the research and graduate education enterprise has grown rapidly and a number of high-profile cases of irresponsible behavior have emerged (*New Scientist* 2012). To address these issues, the Peking University Health Science Center, for example, has introduced RCR education for graduate students and is taking other steps to promote research integrity at the undergraduate level (Cong 2013). Training courses include both online and face-to-face learning, with discussion of cases as an important component.

Many institutions, and even entire countries, may try to downplay instances of misconduct to avoid negative publicity. But institutions that deal forthrightly and openly with problems generally fare better than those that try to cover them up. Overcoming a culture of face saving can make an institution and the entire research enterprise stronger.

Research institutions also need clear, well-communicated rules that define irresponsible conduct and ensure that all researchers, research staff, and students are trained in the application of these rules to research. They also need impartial and confidential mechanisms to investigate suspected breaches of the rules. Each institution should have a standing committee that deals with such cases, or it should establish an ad hoc committee when serious allegations are made. The process should be divided into two stages: a preliminary inquiry to establish whether there is credible evidence that the rules have been breached, and a careful investigation when this is warranted. Investigations should take place as quickly as possible, and the response to findings of irresponsible actions should ensure that the research record is correct. Institutions may find it helpful to incorporate faculty from other institutions into investigation committees to counteract possible bias.

As an example, research institutions and major research sponsors in the United Kingdom issued the *Concordat to support research integrity* (UUK 2012). The concordat spells out commitments to "maintaining the highest standards of rigour and integrity in all aspects of research," and to "ensuring that research is conducted according to appropriate ethical, legal and professional frameworks, obligations and standards," among others (see box 6-1).

Box 6-1. Focus: Commitments of Research Employers in the United Kingdom to Dealing with Allegation of Research Misconduct

In Commitment 4 of The *Concordat to support research integrity* (UUK 2012), signatories pledge to use "transparent, robust and fair processes to deal with allegations of research misconduct should they arise." Universities and other employers of researchers specifically commit to:

- Have clear, well-articulated, and confidential mechanisms for reporting allegations of research misconduct

- Have robust, transparent, and fair processes for dealing with allegations of misconduct that reflect best practice

- Ensure that all researchers are made aware of the relevant contacts and procedures for making allegations

- Act with no detriment to whistle-blowers making allegations of misconduct in good faith

- Provide information on investigations of research misconduct to funders of research and professional and/or statutory bodies as required by their conditions of grant and other legal, professional, and statutory obligations
- Support their researchers in providing appropriate information to professional and/or statutory bodies

---

Some countries have had good experiences with independent ombudsmen who can handle issues of irresponsible conduct both on an institutional level and on a national level. Ombudsmen generally do not have the power to initiate investigations, but they should be able to require institutional or independent investigations of suspected irresponsible practices. Other institutions have designated offices or individuals responsible for hearing allegations and determining the appropriate course of action. Procedures differ among organizations and among countries, but all researchers need multiple options within a research institution that they can pursue if they have witnessed or suspect irresponsible practices. Sanctions against researchers who have violated policies also need to follow employment law in the country where they are working.

In responding to reports of irresponsible practices in research, some principles should be universally observed. Researchers who raise concerns about irresponsible research practices must be protected. Responding to such practices can put a whistleblower's career at risk and is never easy (Gunsalus 1998). All researchers must be confident that they can take action without incurring reprisals.

At the same time, those accused of irresponsible practices need to be treated fairly. Due process, proper communication during an investigation, and fair adjudication are essential. Humans are fallible, which means that accusations of irresponsible actions may be mistaken or malicious. The groups handling such accusations have a heavy responsibility. Freedom of belief, research, and speech must be accorded equally to the accused and the accuser. Box 6-2 illustrates some of the complexities that can arise for institutions when allegations of irresponsible research are investigated and addressed.

Box 6-2. Discussion Scenario: Institutional Response to an Allegation

You are the director of a research institute focused on cell biology. A year after publishing a paper on cell signaling in a prestigious journal, one of the institute's most promising graduate students submits a manuscript to the same journal. One of the other graduate students in his lab noticed that the gel images of control proteins in the newly submitted manuscript are identical to images in the previously published paper, and she reports her suspicion to you. You form an institutional committee to assess the situation. The authoring graduate student states that although he may have mislabeled some images and mixed up which control images belonged with which data sets, this did not affect the experimental results. Your committee agrees to reprimand the student and instructs the lab's PI to retract the submission. Although the PI agreed to implement these instructions, you find out that he disregarded the agreement when the article is published. The PI explains that the student who had called attention to the images had done so out of malice—she had been romantically involved with the student under investigation, and they had broken up on bad terms. The journal has contacted you after several readers quickly noticed similarities in the images between the two papers. Your national government, which sponsors the institute, does not have a formal policy on research misconduct, but the media have picked up the story because the PI is nationally prominent. The authoring student has now graduated and is a postdoctoral fellow in another country.

What are your next steps, both internally (with regard to the PI) and externally (with regard to the journal and the media)? What lessons will you take from this incident regarding procedures for dealing with allegations of undesirable research practices and regarding education and training?

Responsibilities of Research Funders and Governments

Experience indicates that a full solution to some problems cannot depend only on research institutions but also requires an independent organization that researchers can contact to discuss concerns. This independent point of contact may be provided by professional societies, by government, or by some other organization. Training of researchers needs to include information about these options.

Funding agencies, whether public or private, have the power to insist on responsible research practices on the part of grantees. They therefore have the right to insist upon the application of appropriate and transparent rules of research conduct. Researchers, in turn, are responsible for the proper handling of the funds entrusted to them. As in other areas, the policies and approaches of funding agencies and other responsible government entities around the world can vary and change over time. For example, some funding agencies mandate that students receive training in responsible conduct of research, while many others do not.

Funding agencies and governments may also set standards and definitions for research integrity. For example, some irresponsible behaviors may be identified as "misconduct" or "fraud." Fabrication of data, falsification of data, and plagiarism are defined as misconduct in a number of places. Some funding agencies and governments also include other irresponsible behavior in this category, such as interference in a research misconduct investigation, misrepresentation, or breach of duty or care. At the same time, numerous funding agencies and governments have not set such standards or definitions.

Public funding agencies or other government entities may also oversee the investigations of research institutions or have the capability to perform their own. Box 6-3 provides a typology of some of the most common national approaches. This diversity in national approaches raises the question of whether greater international harmonization in policies and standards is desirable or achievable, which is taken up in more detail in chapter 9.

When developing requirements for research institutions in the area of research integrity, governments need to be sensitive to the administrative and other costs borne by institutions in complying with these requirements. The

Box 6-3. Focus: Diversity in National Approaches to Ensuring Research Integrity

The 2010 report of the Council of Canadian Academies *Honesty, Accountability, and Trust: Fostering Research Integrity in Canada, Report of the Expert Panel on Research Integrity* presents a typology of national approaches. These approaches fall into several categories:

Type I consists of nationally legislated, centralized systems with investigatory powers

Includes countries that have established national bodies to investigate and report upon research misconduct as defined by the country in question, generally responding to notifications from institutions or allegations brought by other parties.

Type II consists of nonlegislated bodies that defer to granting agencies or individual institutions for oversight

Includes countries whose national systems have not been established through legislation but whose policies define misconduct and establish guidelines for addressing it.

Type III consists of systems that lack an independent research integrity oversight body or compliance mechanism

Includes countries that do not have a national research integrity oversight body.

imposition of unfunded mandates can damage research productivity.

New technologies and the digitization of past work raise the possibility that irresponsible behavior from many years in the past may be uncovered. For example, politicians and other prominent people have been found to have plagiarized parts of their doctoral dissertations or misrepresented their credentials years after the fact and have suffered serious consequences (Times Higher Education 2013; *New York Times* 2014). Should there be, in effect, a statute of limitations for irresponsible behavior? The research record should be corrected if it contains errors arising from irresponsible actions. However, the issue of sanctions or punishment will depend on the laws and regulations of individual countries. In addition, the degree of forgiveness extended to public figures for past missteps that are uncovered varies according to the country and the circumstances.

Responsibilities of Academies and Interacademy Organizations

Academies and interacademy organizations need to provide forceful leadership on matters of research conduct. They should help to establish standards for the responsible conduct of research and should play an active role in disseminating those standards. This should include communication with younger researchers, including academies for young researchers where these exist. Academies should work within their own scientific communities to ensure that effective mechanisms exist to address allegations of research misconduct. At a regional and global level, interacademy organizations can play analogous roles.

Academies that manage research institutes bear the responsibility for creating a culture of research integrity and dealing properly with allegations of irresponsible conduct. Other academies have a standing committee on research ethics with an advisory function. Some academies have responsibility for investigating allegations of misconduct among their fellows or acting as an advisory "supreme court" in cases of remaining discontent.

Most academies do not have the capacity to investigate cases of alleged misconduct, reach a verdict, or make recommendations for punishment. Nor do academies have the legal authority to serve as a court of appeal where either the accused or the accuser can lodge an appeal against a decision. However, academies can serve in an advisory role for other organizations in difficult or complicated cases. Academies also can monitor issues involving research conduct and reflect on the basic norms and standards in science and scholarship and on the prevalence, causes, and possible ways of preventing breaches of research integrity. This reflective role can be supported by analyses of the literature, reports of work groups, and conferences.

7

ALIGNING INCENTIVES WITH RESPONSIBLE RESEARCH

As the research enterprise has grown, an increasing number of countries and research institutions are facing the challenge of supporting and undertaking research that advances new knowledge and supports other national goals while being conducted at high levels of quality and integrity. Changes in the research system, including opportunities for commercialization, changing institutional environments, and funding pressures, can create incentives that either encourage responsible conduct in research or heighten temptations to violate standards. All participants in research need to understand these environmental factors and work to promote research integrity by managing, mitigating, or eliminating potentially harmful incentives.

While beginning researchers and their supervisors are a primary audience for this guide, this chapter describes the responsibilities of other individuals and institutions involved in research. All researchers will have interactions with research institutions throughout their careers, and all researchers will have opportunities to influence the policies and actions of these institutions.

Managing Individual and Institutional Conflicts of Interest

Researchers have many different interests in their professional and personal lives, and sometimes these interests conflict. For example, a researcher may have a financial interest that inappropriately biases his or her perspective on a research program. Or a researcher asked to serve as a reviewer may have a personal conflict that would affect the integrity of the review. Conflicts of interest can be real or perceived, but both can tarnish a project and the researchers associated with that project if they are not disclosed and managed.

Many conflicts of interest arise from opportunities to capitalize on research findings by securing intellectual property protection through patents, copyrights, or trade secrets and then licensing this knowledge or forming a start-up company to exploit it. Many countries treat the commercialization of research results as a legitimate and valuable activity. Researchers and research institutions should be aware of their rights and responsibilities under national laws and funder policies. The U.S. National Academies publication *On Being a Scientist* has a succinct discussion of intellectual property issues related to research as they are encountered in the U.S. context (NAS-NAE-IOM 2009).

Financial conflicts such as research support, consultancies, stock ownership, honoraria, or other payments are fairly easy to identify. However, personal relationships, competition among researchers, or strongly held intellectual views also can create conflicts. In addition, researchers may put themselves in situations where ostensibly part-time activities such as consulting demand so much of their time and energy that their core research responsibilities suffer—a circumstance known as a *conflict of commitment*.

Researchers are responsible for disclosing upon request all the financial and personal relationships that might bias their work (see box 7-1). Research institutions, research sponsors, and journals are increasingly requiring researchers to disclose conflicts of interest. For example, when submitting a manuscript to a journal, researchers typically have to state explicitly whether conflicts exist. The International Committee of Medical Journal Editors has prepared a uniform disclosure form that journals can use or adapt for conflict of interest disclosures (ICMJE 2013). Researchers also have a responsibility not to enter into agreements that would interfere with the disclosure of biases.

Box 7-1. Discussion Scenario: A Personal Conflict of Interest

You are a graduate student completing your PhD dissertation and are invited to peer-review a manuscript for a journal for the first time. The peer-review system is a hallmark of the scientific process and you are excited to be part of it. You read the abstract and believe that your expertise allows you to perform a thorough review and accept the invitation to receive the full manuscript. While reading the paper, you are able to deduce that the first author is a close personal friend with whom you worked in the past and who will soon be looking for a tenure-track position. You also notice that the paper contains significant flaws in the data-analysis section, and you believe that it should be substantially revised or rejected for that reason.

What would you do in this situation? How would you disclose the conflict of interest? What are the implications of not disclosing your conflict in this situation?

Research institutions can have financial, reputational, and other conflicts that might affect their oversight of research or willingness to impartially investigate suspected irresponsible behavior. One well-known example is the case of Jesse Gelsinger, who died in a gene therapy clinical trial in 1999 conducted by the University of Pennsylvania (Steinbrook 2008). Gelsinger was substituted for another participant in the trial despite his having a condition that should have excluded him. In addition, the university did not report that other patients had experienced serious side effects from the treatment. The PI had founded a company to commercialize the technology, and the university held a stake and received research funding from the company.

Corporate sponsorship of academic research activities can deliver significant benefits, such as exposing faculty and students to interesting practical problems and introducing students to career possibilities in industry. At the same time, academic research institutions and researchers need to ensure that corporations and other sponsors do not exert inappropriate influence over research and educational activities. For example, sponsored research agreements that place onerous restrictions on performing research or reporting results can impair academic freedom and damage the educational process and faculty careers. In some cases, relationships between corporate interests and academia have been even more destructive (Oreskes and Conway

2011). An extreme example is the tobacco industry, which used research funding as one tool to obscure the unhealthy effects of smoking for many years (Proctor 2011).

Research Institutions

In addition to their obligation to establish and enforce codes of conduct, research institutions have a broader responsibility for maintaining an environment that fosters research integrity (NRC-IOM 2002). Researchers, research institutions, and research sponsors have learned that no country or field is immune from irresponsible actions. The fundamental values of research need to be practiced and emphasized as a matter of routine. Experienced researchers need to convey the standards of research through teaching, through the examples they set, and through mentoring to students and younger colleagues. A growing number of universities are taking steps to assess and improve the research integrity climate of individual departments and the institution as a whole through a standard survey and follow-up activities (CGS 2012).

Institutions that employ researchers thrive when they emphasize excellence and creativity. In recent years, hiring, promotion, and funding decisions have made increased use of such metrics as the number of citations a publication has received or the "impact factor" of a journal calculated from citation of articles in that journal. However, overreliance on such metrics can be misleading and can distort incentive systems in research in harmful ways. For example, researchers may try to publish as many articles as possible, reducing the quality of their articles as a result. The provision of large bonuses to researchers who publish a paper in a prestigious journal may provide a temptation for researchers to cut corners or worse.

The value of a researcher's contribution cannot be measured solely by the number of publications or the prestige of the journals that publish them. In 2014, the San Francisco Declaration on Research Assessment (DORA) emphasized the need for new more holistic ways to evaluate research achievements and avoid overreliance on metrics such as citation indices and impact factors.

Research Funders

The public and private agencies that support research, including governments, philanthropies, and industry, need to support the best research possible. **Funding agencies should provide support to researchers and research institutions at a level sufficient to ensure that research can be undertaken properly and responsibly, without compromising quality or integrity.** Their funding policies should not promote an environment where researchers face strong incentives to publish as many papers as possible in a short period of time or otherwise compromise the quality or integrity of their research.

As the global research enterprise has grown and expanded, competition among researchers for recognition, positions, and limited resources has become more intense. In some fields and in some countries, growth in funding has been flat as the number of researchers has grown, leading to a situation in which a smaller proportion of proposals are funded and a smaller proportion of graduate students and postdoctoral fellows can expect to achieve an independent research position (Stephan 2012). Such structural factors may be partly responsible for the persistence of irresponsible practices and other negative impacts on science (Casadevall and Fang 2012).

Funders have the corresponding duty to provide funding sufficient to ensure that researchers and research institutions can put systems in place that uphold integrity and facilitate high-quality research. In particular, funding agencies should support efforts of research institutions to develop education and training programs on responsible research conduct. They should refrain from awarding research funding on the basis of inappropriate political bias. Also, unless a researcher has signed a contract imposing limits on publication, that researcher has the right to publish research results without constraints from funders.

Because the results of research can be difficult to predict, funders often must give researchers considerable latitude in deciding which questions to pursue and how to pursue them. Early-career researchers in particular need both independence and support to establish their careers while following their passions and interests.

8

REPORTING RESEARCH RESULTS

ETHICS ON THE EDGE

I see you're the co-author of this paper, Dr. Mauritz, and you came up with some new insights in the field of quantum mechanics, which you will explain further next week.

I AM?
I DID?
I WILL?

Researchers have many ways of reporting research results to others. They can discuss results within a research group, give a presentation at a meeting, prepare a poster for a conference, write about results in a blog or other electronic forum, construct a database and make it available electronically, or publish a journal article, a chapter of a book, or a book. All these types of communications need to observe the essential values of honesty, fairness, and openness. Many kinds of irresponsible and undesirable practices are associated with the reporting of research results.

Authorship

Authorship serves to identify the individuals who have made substantial intellectual contributions to a study or are responsible for a component of the work behind it. Authorship underlies the reward systems of many research institutions and also entails accountability for published products.

Authorship criteria and conventions such as the order in which authors are listed differ by discipline. The statement on authorship by the International Committee of Medical Journal Editors (ICMJE 2015), for example, is especially demanding. It sets forward these authorship criteria:

- Substantial contributions to the conception or design of the work; or the acquisition, analysis, or interpretation of data for the work; AND
- Drafting the work or revising it critically for important intellectual content; AND
- Final approval of the version to be published; AND
- Agreement to be accountable for all aspects of the work in ensuring that questions related to the accuracy or integrity of any part of the work are appropriately investigated and resolved.

However, most accepted guidelines dictate that authors need only to provide substantial contribution to the manuscript in one or more of the previously mentioned criteria.

Within particular disciplines, the order of the authors in an article traditionally conveys information about the roles of those authors in the research being described. But this inferred information can be ambiguous or misleading, and the meaning of author order in interdisciplinary research can be indecipherable or meaningless. Authorship also differs from country to country; while the PI is listed as the last author in some countries, in others the PI is listed first.

Developing strict, multidisciplinary guidelines about author order in cases where the order denotes the importance of contributions would be difficult, if not impossible. Some journals or research institutes try to avoid these difficulties by listing authors alphabetically. Others require that the roles of authors be described in a note accompanying the article. The drawback of such notes is that it can act to absolve some authors of responsibility for the article if they can point to a note stating that their roles were limited.

Unless a paper specifically allocates responsibility among authors, authorship connotes responsibility for the entire contents of that paper. The authors of a discredited paper may claim that they do not have expertise in the part of a paper containing fraudulent or erroneous results, especially in multidisciplinary research. However, if a paper contains fraudulent or erroneous results, all authors will be held accountable for those results. An author without expertise in a particular area may need to ask a trusted colleague to review a paper to have confidence in its accuracy.

Sometimes, the authors of a paper add an author who has not contributed to the paper to honor the author, to boost the paper's visibility, or to increase the chances that the paper will be accepted by a prominent journal. In other cases, senior researchers demand to be listed as authors even though they have not contributed to the paper (box 8-1). Hierarchical pressures in research organizations may lead authors to list laboratory or institute directors who have not contributed. This issue can be particularly problematic for beginning researchers, who may then face the difficult choice about whether to share their concerns with institutional officials or take other actions (Gunsalus 1998). Sometimes, papers are written by authors who had nothing to do with the research described, and a researcher's name is affixed to the paper in lieu of the actual writer. Both "guest authors" and unacknowledged "ghost authors" undermine the standards of research and distort the allocation of credit.

Box 8-1. Discussion Scenario: Honorary Authorship

You are a professor who recently received tenure at one of the leading research universities in your home country after earning your PhD in another country. You are very excited about the results of recent experiments, which are significant enough to merit publication in a leading international journal. As you complete work on the manuscript for submission to one such journal, your department chair points out that acceptance of your paper will result in large financial bonuses for you and your coauthors personally and lead to a significant funding increase for the department. He suggests that you add your graduate advisor at the overseas university, who was not involved with the research but is internationally known in the field, as a coauthor on the paper. This would surely improve the odds that the paper will be accepted. The department chair also indicates that he expects to be a coauthor on the paper as well, even though he has not been involved with the work.

How would you respond to the department chair? What possible consequences can you foresee if you follow his suggestions?

Determining authorship in the early stages of a research project may not be practical. In many cases it is impossible to predict what balance of efforts will produce the final published work. However, it is important for authorship to be discussed and for all collaborators to understand the criteria that will be used to determine authorship. For example, all researchers involved in a project may agree on a set of principles that will be used to determine who is an author and the order of authors once the paper is ready for submission. Research institutions can also create mechanisms to ensure that those principles are respected and act in case of conflict.

Peer Review

In addition to judging the merit of research proposals, peer review is used to judge the merit of communications submitted for publication as well as to improve those articles through constructive criticism. Peer review seeks to ensure that the communication is relevant, that the evidence supports the conclusions, and that the findings are of value. It can enhance the quality of publications by clarifying explanations, correcting errors, properly allocating credit, and enabling other improvements. Publishing in journals and with publishers known for their high standards of peer review enhances the reputation of authors.

Peer review of proposed publications can take several forms. The most common arrangement is for the reviewer to be anonymous to encourage honest and frank reviews, and most reviewers favor this approach. Another approach is for both the authors and the reviewers to be blinded, although authors and reviewers can sometimes surmise the identities of reviewers or authors. A third method is for the entire process to be open, with the reviewers and authors both identified and the comments from both sides made freely available. In a fourth method, either before or after publication all readers and reviewers can access the publication and provide comments, generally in an online forum. Also, many journals have added electronic forums where readers can post comments on a published article. Especially common in the life sciences, this practice has not yet become formal or institutionalized enough to provide a replacement for peer review. At this point, it is not clear what the future balance might be between traditional peer review and alternatives.

Irresponsible practices in peer review can occur when the reviewer is biased for or against the authors or has competing interests. To minimize such conflicts, some journals allow authors to name persons to whom an article should not be sent for peer review. If an article is rejected, some editors allow the authors to appeal the decision. **Peer reviewers need to assess proposed publications fairly and promptly, with full disclosure of conflicts of interest or bias.**

Some authors have complained that a publication has been kept on hold for an unnecessarily long time while a reviewer finishes a competing publication. Such fears can be especially keen for authors who are at a disadvantage in peer review, including researchers from countries that are not at the center of a research field. Some authors also have complained about racial or gender discrimination in review decisions. Potential reviewers who realize that they have a conflict, a bias, or a lack of needed background knowledge in reviewing a proposed publication have a duty to inform editors so that appropriate actions can be taken.

Peer review sometimes detects fraudulent research, and reviewers should report any anomalies or suspicions to the editors, but reviewers generally must trust that the work described was done honestly. Peer review is not designed primarily to detect irresponsible practices, such as using public data as if it were the author's own, submitting papers with the same content to different journals, or submitting an article that has already been published in another language without reference to the original.

Difficulties can arise in reviewing publications from large collaborative projects involving researchers from different institutions, different countries, or different research disciplines. Reviewers may need to be comparably diverse to judge the multiple aspects of such a publication. Another area of concern has been the communication of dual-use research, such as research results that could contribute to the development of chemical or biological weapons. In such circumstances, reviewers or specially constituted panels may be asked to determine whether the likely benefits of publication outweigh the possible risks.

Peer review also faces a long-term challenge stemming from the fact that in most cases it constitutes volunteer labor on the part of the reviewer. Given that research has become a faster-paced and more competitive activity, researchers may become less inclined to undertake service activities such as reviewing. There are ways that this could be addressed, such as by paying reviewers. However, it is not obvious where the resources would come from to implement this approach.

Abuses of Publication Practices

Many other forms of irresponsible conduct can be associated with publication practices.

The citations in a paper acknowledge the previous work on which research results are based. Researchers who fail to acknowledge the contributions of others place their reputations at risk.

As discussed in chapter 4, the practice known as duplication—publishing the same research in multiple places—wastes the time and other resources of reviewers and editors and is fundamentally dishonest when a reader expects a research report to be original. In contrast, publication of the same work in multiple languages can be valuable, particularly in cases where the original work appears in a less widely spoken language. The circumstances of such republication should be made clear to editors at the time of submission and to readers. The obligation to appropriately cite the source literature applies regardless of the language in which a paper is published.

Publishing different results from a research project in as many places as possible to increase publication counts—a practice known as salami slicing or publishing in "least publishable units"—does a disservice to readers and editors who would benefit from a more thorough treatment of the research. This practice, too, wastes the time and effort of others and renders the research literature less useful than it would otherwise be.

Editors should refrain from encouraging or coercing authors to add citations from their journal in order to boost the journal's impact factor. This practice distorts the traditional mechanisms for judging the importance and relevance of results.

Journals, research institutions, and individual researchers can fall prey to the temptation to enhance public recognition of results by engaging in such irresponsible behaviors as inaccurate representation of the results' implications or rushing the publication process. In 2013, a highly publicized paper on creating human stem cell lines through cloning was accepted three days after submission and published twelve days after that (Tachibana et al. 2013a). However, the paper was quickly found to contain image duplication and labeling errors, resulting in concerns over manuscript preparation and the review process (Cyranoski 2013). One month later, an erratum was published to correct figures and typographical errors (Tachibana et al. 2013b), and the authors promptly shared all reagents for independent validation (Chung et al. 2014; Yamada et al. 2014). This case raises questions over whether groundbreaking results should be replicated before or after publication and highlights the importance of providing clear descriptions of any image manipulation (Cyranoski 2014).

Finally, the use of computer programs and large publicly accessible databases to do text and data mining of research results has opened up grey areas of research practice that need further investigation. For example, should researchers be able to generate commercially valuable products from open access databases without financial return, or in some cases even attribution, to the researchers who gathered, analyzed, and publicly disseminated those data?

The Role of Journals

As repositories of the research literature, journals have a responsibility to maintain the integrity of research results. This entails establishing not only proper peer review processes but also proper handling of retractions. When a published paper is shown to be based on fraudulent data, journals have a responsibility to issue a correction or retract the paper. Although journals should make retractions and corrections visible in print and electronic versions so that retracted papers are no longer cited and corrected versions are used, many fail to do so. Also, journals may be reluctant to communicate whether a retraction was the result of an honest error or irresponsible conduct (Fang et al. 2012), sometimes because national laws prohibit potential libel of authors.

Maintaining the integrity of the research literature requires more than peer review and proper handling of retractions (see box 8-2). An increasing number of journals are using software or manual screens to guard against plagiarism and the inappropriate manipulation of figures.

Box 8-2. Focus: Irresponsible Behavior in Stem Cell Research

Stem cell research has been a major focus of the life sciences for over a decade and has seen several prominent cases of irresponsible research behavior. The case of Hwan Woo-suk, a researcher at Seoul National University in Korea, attracted widespread attention in 2005 and 2006 when it was discovered that he used falsified data and committed other transgressions in two papers published in *Science* (*Science* 2006).

A more recent case involves the RIKEN Center for Developmental Biology in Kobe, Japan (*The Guardian* 2015). In January 2014, two papers were published in the journal *Nature* to describe a simple method to create stem cells after exposing blood cells from mice to an acidic solution (Obokata et al. 2014a, 2014b). A few months later, the journal discovered plagiarized writing, manipulation of images, and inexplicable discrepancies in the reported data. Haruko Obokata, lead author of both manuscripts, was found guilty of falsifying data. Her supervisor, Yoshiki Sasai, who joined the project in the final writing stage of the manuscript, was severely criticized during the investigation process for his poor mentoring and committed suicide at the research institution a few weeks after the journal retracted both studies. This case illustrates the human toll that irresponsible research behavior can take.

If reviewers raise concerns or journals detect issues after review, the editors may communicate with the author to determine whether an error was accidental or the result of irresponsible practices, and they may ask the authors for the raw data on which a conclusion is based. If evidence of misconduct surfaces, a journal should inform an author's institution of the infraction, but this practice is not universal. The Committee on Publication Ethics (COPE 2011) has established a code of conduct and retraction guidelines and provides advice to editors and publishers on publication ethics.

Changes in Scholarly Communication

A number of shifts are occurring in scholarly communication that affect the environment for research integrity. One such trend is the requirement by public and private funders that the results of research be open to all. This requirement, which has become most widespread in biomedical fields, often involves an embargo period when access is restricted to journal subscribers. After the embargo period, papers are posted in an online repository.

Open access challenges the business model of traditional journals, which is based on income from subscribers. Traditional journals add value to the publishing process, and they must be economically viable to exist. Yet to the extent that research results are freely and widely available, they increase in public value. Whether and how this tension will be resolved is not yet clear. New journals based on an "author pays" model have sprung up, and some have become very successful and influential.

Open access also has enabled the emergence of what University of Colorado in Denver librarian Jeffrey Beall refers to as "predatory open-access journals" that accept papers in return for payment from authors but apparently undertake no quality control (Bohannon 2013; Kolata 2013). For example, a science journalist fabricated a paper about the discovery of an anticancer drug and submitted 304 versions of it to open-access journals (Bohannon 2013). Surprisingly, more than half of the journals accepted the manuscript, not noticing that the results were meaningless and that the author and institution were fabricated. For senior researchers with experience publishing scholarly papers, it is quite common to receive email invitations to submit papers to newly created journals under appealing names. Researchers should be wary about publishing in such journals, as it could amount to a waste of their creativity and hard work, and these practices can seriously damage the scientific enterprise. There is an ongoing debate within the scientific community about how to use open-access practices for the betterment of science (OSI 2015).

In some fields, the release of research results before they have been peer reviewed is frowned upon, and some journals will not allow distribution of "preprints." In other fields, the availability of preprints is seen as speeding progress.

One well-established example is the arXiv, a web archive of electronic preprints in mathematics, physics, astronomy, and other fields that has reached the milestone of one million papers (arXiv 2015). The site has instituted practices aimed at ensuring that submissions are high quality and of interest to the community, including the use of subject matter moderators and an endorsement system.

Websites and blogs that investigate or report on scientific publishing issues are also having an impact. Perhaps the most influential of these is Retraction Watch (2015). The blog has increased pressure on publishers to explain the reasons for retractions and has allowed the community to more efficiently track cases in which a given researcher is responsible for large numbers of retracted papers.

9

BENEFITS AND CHALLENGES OF INTERNATIONAL COLLABORATIONS

"WHAT IF WE SPEND ALL THESE BILLIONS, AND THERE JUST AREN'T ANY MORE PARTICLES TO FIND?"

As the global research enterprise grows and diversifies, an increasingly number of researchers are crossing national borders to pursue education, research opportunities, and careers. These international collaborations have many benefits for individual researchers, for their institutions, and for nations. Researchers gain access to ideas, facilities, and new experiences while building relationships that can last a lifetime. Collaborations can build political and economic links that transcend boundaries. Indeed, research collaboration forms an important component of foreign policy for a number of nations. Large multinational projects share the cost of science while accelerating the production of knowledge that benefits all countries. A recent analysis indicates that internationally coauthored work is more highly cited than collaborative work within one country (Adams 2013).

International experiences can be a high point of a person's career in research. But international research also can raise issues of responsible conduct that do not arise in a purely national context.

Special Challenges in International Collaboration

International research can take many forms. A research group may include someone from another country, or a student or postdoctoral fellow may seek research opportunities in another country. PIs from two different countries may collaborate, or investigators may belong to international networks of researchers working collaboratively. Research institutions may be linked, or an institution may have a satellite campus or subsidiary in another country. Some research problems are too large and complex for any one country to tackle, requiring that large infrastructures be established to make progress on them. Box 9-1 shows one typology of international collaborations.

Box 9-1. Focus: Modes and Drivers Behind Different Types of International Research Collaborations

According to Tony Mayer, Europe Representative of Nanyang Technological University, collaborations are driven by a need for more international interdisciplinary and multidisciplinary

---

knowledge. All require trust and integrity (Mayer 2013). He suggests the following typology:

- **Classical Mode:** Collaboration between two PIs
- **International networks of PIs** (example: European Cooperation in Science and Technology-COST)
- **Networks of institutions** (example: GlobalTech, the Global Alliance of Technological Universities)
- **Satellite campuses or subsidiaries in other countries** (example: the National Research Foundation of Singapore's Campus for Research Excellence and Technological Enterprise—CREATE—program, which supports universities setting up research programs in Singapore)
- **Connections via super infrastructures** (examples: CERN, ITER, Integrated Ocean Drilling Program)
- **Research on global challenges such as climate change** (Earth System Science)

The breadth and range of international research collaborations are very broad, so only a few illustrative examples will be mentioned here. Some international collaborative research programs, such as the network of agricultural research center known as CGIAR (originally the Consultative Group on International Agricultural Research), have existed and evolved over many years. A newly launched initiative is the Institute for International Crop Improvement, whose main focus is to improve crops that are "important for food security in developing countries" (Danforth Center 2015).

Another example of international collaboration is the Census of Marine Life, completed in 2010. This ten-year international effort involved two thousand, seven hundred scientists from approximately eighty nations to assess the diversity, distribution, and abundance of marine life at the global level. This project produced the most comprehensive inventory of marine life to date, providing unmatched data for forecasting, measuring, and understanding changes in the marine ecosystem and for building individual, institutional, national, and regional capacity for marine sciences (CoML 2010).

With research involving international collaborators, legal, social, and cultural differences may lead to disputes over whether someone has acted irresponsibly (NRC 2014). In addition, as at the national level, science can have undesirable consequences when misused in an international context. In some countries the freedom of scientists is circumscribed, scientific integrity can be violated, and scientists can put themselves in danger by publishing certain results.

For example, one risk of international research collaboration or the international movement of students is the possibility that a collaborator or student might commit espionage or use the knowledge that they gain for nefarious purposes. Pakistani nuclear scientist A. Q. Khan, who received much of his education and spent his early career at several institutions in Europe, is one cautionary example. Khan returned to Pakistan, ultimately leading Pakistan's successful nuclear weapons program and playing a key role in proliferating nuclear technology to North Korea, Libya, and Iran. Certainly, researchers and institutions need to make their best efforts to vet collaborators and students and remain alert to possible problems. Consulting with experienced researchers and experts in other areas such as the law may be necessary. At the same time, it may be very difficult or impossible to predict the attitudes and actions of an individual researcher decades in advance. A proper balance between caution on the one hand and an appreciation for academic freedom and international collaboration should be maintained. The Royal

Netherlands Academy of Arts and Sciences published a brochure on this sensitive subject that lays out a framework for the analysis of the challenges and predicaments of international scientific cooperation and offers options for risk assessment (KNAW 2014).

When an individual researcher or research team invites a foreign colleague, postdoctoral fellow, or student to participate in the research, the rules of conduct of the country where the research is being undertaken normally apply. Host and guest researchers should be thoroughly familiar with these rules and agree to abide by them.

When two or more researchers or groups of researchers from different countries decide to work together on a research project, national codes or procedures may be at variance or even contradict each other. Under these circumstances, the codes and procedures to be followed need to be specified before the start of the collaboration. Potential problems such as dual-use technologies, intellectual property, and sharing of research materials should be discussed beforehand. The criteria and process for determining authorship should be established along with agreements on how to share raw or processed data. Experimental procedures should be adapted to the available infrastructures, and materials produced as part of the collaboration should avoid nonscientific statements and be peer reviewed. When supporting international research collaborations, funding agencies should make sure that rules are clear and understood by all parties to the collaboration in advance.

The European Code of Conduct (ESF-ALLEA 2011) recommends that international collaborations follow the guidance of the Organization for Economic Cooperation and Development Global Science Forum (OECD 2007). The forum has produced representative text for international agreements that can be embodied or adapted in documents for collaborative projects.

Box 9-2. Discussion Scenario: Miscommunication in an International Collaboration

You are a graduate student who has qualified for a four-month research fellowship from a foreign government that will cover all your expenses. To receive the award, you need to prepare a short research proposal and find a research team in the foreign country willing to host you.

You are enthusiastic about the opportunity, but your capabilities in English are limited and you have no proficiency in the language of the country you will be visiting. You struggle to communicate with potential advisors. So far, you have not received any outright rejections, but the responses have been vague: for example, "we will look into that," or "it seems like a great project, we will think about it." Time is running out, and you need to receive formal confirmation from an institution. However, you feel uncertain about how to let the potential advisors know that you need a clear response about whether they are willing to accept you.

Are there steps you could have taken to communicate more effectively? What are the responsibilities of potential advisors when they receive this sort of request from an overseas researcher?

The Need for Harmonization

The growing internationalization of research requires that more attention be devoted to differences in definitions, standards, and procedures among nations. Differences among countries can be especially difficult when two or more countries need to investigate an allegation of irresponsible conduct. Officials and administrators responsible for promoting and enforcing integrity in research need to collaborate just as researchers collaborate.

Smaller and less wealthy countries can face particular difficulties in international collaborations. Alliances of those countries, such as those established through IAP—The Global Network of Science Academies, can help ensure that the concerns of these countries are represented in international deliberations.

The *Singapore Statement* that emerged from the Second World Conference on Research Integrity (2010) was designed to be a global guide to the responsible conduct of research. The Montreal Statement of the Third World Conference on Research Integrity (2013) outlines principles to be followed in organizing and undertaking international collaborations.

Greater harmonization of training programs could reduce differences in perceptions of codes of conduct. In particular, online collaboration among institutions could help harmonize education programs (Steneck 2013).

10

COMMUNICATING WITH POLICYMAKERS AND THE PUBLIC

"WE SHOULD MENTION A FEW MILD SIDE EFFECTS. IF THERE ARE NONE AT ALL, PEOPLE WILL BE SUSPICIOUS."

Many people outside the research enterprise are interested in and use research results. Government officials may draw on the results of research to make regulatory decisions. Public policies in a wide variety of areas incorporate scientific information. Discussion of prominent issues in the media may hinge on information from research. Even the results of fundamental science are of interest to a public that has grown up in an age of rapid scientific and technological advances.

Researchers are often asked to communicate information to policymakers and the public, and several principles of responsible conduct govern these interactions.

Participating in Public Decision Making

Researchers have much to contribute to public policies in a wide variety of areas. But research-based evidence is not the only input to such policies. Among the many factors that can influence decision makers are stakeholder input, political convictions, election cycles, media coverage, popular support, competing issues, staff input, and lobbying. Policy makers often rely on trusted individuals to provide them with good advice. They can be more influenced by a powerful anecdote or personal story than by a sheaf of research results.

Evaluating what roles are appropriate for researchers in public processes can be complex and nuanced. It may depend on the national policy context, the nature of the issue, the current state of science in relevant areas, and other factors. When providing information for public policy decisions, one possible role of the researcher is to be an honest broker. Researchers are particularly well suited to untangle basic scientific facts from other considerations. However, researchers need to make their inputs both useable and transparent for decision makers. The users of research results need to be able to judge the reliability of the evidence, of the claims that are made, and of those making the claims.

A particular difficulty is communicating uncertainties or probabilities clearly and comprehensively. Policymakers may want fixed inputs to a decision rather than a range of possibilities. Research results, in contrast, can be ambiguous and uncertain. Researchers need to think about how their input can be most useful despite these constraints. Box 10-1 provides an example of what can go wrong when

Box 10-1. Focus: The L'Aquila Earthquake

Early in the morning of April 6, 2009, 309 people were killed when a 6.3-magnitude earthquake struck the town of L'Aquila, Italy, in the seismically active region of Abruzzo. In October 2012, a judge found 7 men—6 scientists and a government official—guilty of manslaughter for giving a falsely reassuring assessment of risk at a meeting of an official government advisory panel a few days before the quake. They were all sentenced to 6 years in prison and heavy fines.

The International Human Rights Network of Academies and Scholarly Societies along with other groups condemned the verdicts and sentences (IHRN 2012). The convictions and sentences were immediately appealed, and in November 2014 the scientists' convictions were overturned. The government official's sentence was reduced to 2 years.

Before the earthquake, residents in the area become nervous about an increase in seismic activity as a series of small shocks occurred. As a result, a meeting of a subgroup of Italy's National Commission for the Forecast and Prevention of Major Risks was held in L'Aquila (Cartlidge 2012). Following the meeting, several members of the commission held a press conference with local officials. The assessment's finding that earthquakes are impossible to predict is accurate. However, the government officer offered an assurance that "The scientific community tells us there is no danger, because there is an ongoing discharge of energy. The situation looks favorable" (*New Scientist* 2012). This statement and subsequent news reports led residents to believe that a large earthquake was unlikely to happen (*New Scientist* 2012).

The L'Aquila case raises important issues related to the scientific assessment of risks to the public, how and by whom those risks are communicated, and the rights and responsibilities of scientific advisors and public officials. One factor influencing

> the handling of the case is the legal framework for scientific advice in Italy (Nosengo 2012). For example, some countries have established clear rules and regulations governing the provision of scientific advice, and scientific advisors are indemnified from criminal and civil liability. Also, information that emerged during and after the trial raises the question of whether the scientists allowed themselves to be used by public safety officials in what was primarily a public relations operation to reassure the public (Cartlidge 2012). Finally, even granting that the scientists may bear some responsibility for inappropriately reassuring L'Aquila residents, the judge's sentences were startling and more severe than what the prosecutor requested.
>
> The 2011 earthquake, tsunami, and nuclear disaster in Japan is another recent example where the performance of scientists and engineers in protecting public safety has been severely criticized (Normile 2011). The lesson for scientists and governments is to ensure that clear guidance exists for provision of scientific advice, that opinions and recommendations are communicated clearly and accurately, and that rights and responsibilities are clarified.

frameworks for providing and communicating scientific advice are unclear.

Knowing how to make results usable for an audience requires dialog with that audience. Participating in public decision making is a two-way and extended process that requires sharing ideas, experiences, and perspectives over time.

Providing Policy Advice to Decision Makers

Sometimes researchers are asked to provide not just research results but policy advice to decision makers in government, industry, or nongovernmental organizations. This advice can be extremely influential and must avoid bias or parochialism.

Documents generated by researchers to provide advice differ from research articles, but they, too, are based on evidence and reason. Scientific policy advice to governments, industry, or nongovernmental organizations should be peer reviewed to bring the quality control mechanisms of research to bear on that advice. If formal peer review is not possible, informal consultations with peers, including those who would be expected to be critical, may be necessary. An example of scientific policy advice on the international level is the InterAcademy Council's assessment of the management and processes of the Intergovermental Panel on Climate Change, described in box 10-2 (IAC 2010). IAP— The Global Network of Science Academies, the InterAcademy Medical Panel, and their affiliated regional academy networks regularly produce statements on policy issues that are disseminated to policy makers. Another example of internationally coordinated advice is the science-based advisory work of EASAC, the European Academies Science Advisory Council.

Box 10-2. Focus: Assessment of the Intergovermental Panel on Climate Change

The IPCC is a United Nations body that regularly assesses "the most recent scientific, technical and socioeconomic information produced worldwide relevant to the understanding of climate change" (IPCC 2013). The body, which was established in 1988, was awarded the Nobel Peace Prize in 2007.

In 2010, following the discovery of a mistake in IPCC's *Fourth Assessment Report* and the leaked release of emails of climate scientists involved with IPCC, the UN requested that the InterAcademy Council assess the management and report preparation processes of the IPCC. While the IAC panel did not assess the IPCC's scientific findings, the resulting report contained a number of recommendations aimed at strengthening "IPCC's processes and procedures so as to be better able to respond to future challenges and ensure the ongoing quality of its reports" (IAC 2010). The key recommendations (in abbreviated form) were

Recommendation: The IPCC should establish an executive committee to act on its behalf between plenary sessions.

Recommendation: The IPCC should elect an executive director to lead the secretariat and handle day-to-day operations of the organization. The term of this senior scientist should be limited to the time frame of one assessment.

Recommendation: The IPCC should encourage review editors to fully exercise their authority to ensure that reviewers' comments are adequately considered by the authors and that genuine controversies are adequately reflected in the report.

Recommendation: The IPCC should adopt a more targeted and effective process for responding to reviewer comments.

Recommendation: Each working group should use the qualitative level-of-understanding scale in its Summary for Policymakers and Technical Summary, as suggested in IPCC's uncertainty guidance for the Fourth Assessment Report.

Recommendation: Quantitative probabilities (as in the likelihood scale) should be used to describe the probability of well-defined outcomes only when sufficient evidence exists.

Recommendation: The IPCC should complete and implement a communications strategy that emphasizes transparency, rapid and thoughtful responses, and relevance to stakeholders, and that includes guidelines about who can speak on behalf of the IPCC and how to represent the organization appropriately.

Several national academies have launched programs to build bridges between elected officials and scientists in order to improve the science advisory process. For example, the Académie des sciences (French Academy) has set up a pairing program with the French Parliamentary Office for Scientific and Technological Assessment (OPECST). Through this initiative, members of the French Academy and young researchers establish formal dialogue with members of Parliament (MPs). A similar program has been implemented by the Royal Society (UK) to foster ongoing dialogue among MPs, civil servants, and research scientists. These bidirectional initiatives provide MPs with better insights into the scientific enterprise while helping researchers to understand pressures associated with political decision making.

Researchers have the same rights as all other people in expressing their opinions and seeking to influence public policy. But researchers must be especially careful to distinguish their roles as specialists and as advocates. Researchers who choose to be advocates have a special responsibility to themselves and to the research community to be very open and honest about the support for the statements they make. Researchers should resist speaking or writing with the authority of science or scholarship on complex, unresolved topics outside their areas of expertise. Researchers can risk their credibility and the credibility of the research enterprise by distorting their results or otherwise behaving irresponsibly as researchers in support of a policy position, no matter how important that policy position might appear to be.

Communicating Scientific Information to the Public

Talking about research results through the media or directly with the public can be professionally rewarding and personally enjoyable. Members of the public, generally speaking, respect and admire researchers. They are interested in new ideas and in the applications of these ideas. The public supports research both for its practical benefits and for its capacity to educate, entertain, and inspire.

Communicating the results of research to those outside the research community can take time away from research (Baron 2010). But effective communication with policy makers and the public is essential given the pervasive influence of research on the broader society and the potential consequences of miscommunication (see box 10-3).

Box 10-3. Focus: The Long-Term Consequences of Irresponsible Behavior

In 1998, *The Lancet*, a British medical journal, published a research paper that linked the administration of measles, mumps, and rubella (MMR) vaccine with the appearance of autism. Soon after its publication the paper was called into question, with other researchers being unable to reproduce the results. Concerns were also raised about the lead author's undisclosed financial interests. After a thorough investigation, the paper was retracted in 2010. This and other cases of misuse and dangerous "studies have shown" stories are articulately discussed by Goldacre in his bestseller *Bad Science* (2009).

Although results from that study were discredited, the fear of a link between autism and vaccines became widespread in several countries, partly through promotion by some advocacy groups and favorable coverage in some media outlets. The result was a drop in vaccination rates in some places. For example, California experienced a measles outbreak in early 2015 attributable to low vaccination rates (Nierenberg 2015).

This case shows how irresponsible research behavior can have downstream effects for decades. It also demonstrates how important it is for the scientific community to establish when such behavior has occurred and communicate correct information to the public.

Science academies, acting individually and through global and regional networks, can play an important role in enhancing the public understanding of science. For example, IAP—The Global Network of Science Academies, and a number of its members have been active in working to improve science education at the elementary and secondary levels in recent years. One area of focus has been the teaching of evolution, where the IAP's 2006 statement has provided important guidance (IAP 2006).

Conveying complex ideas to a general audience requires the ability to communicate simply and clearly. These skills are not innate in the scientific environment, and researchers may need specialized training in conveying scientific insights to the general public in a way that is helpful and engaging. The American Association for the Advancement of Science's Center for Public Engagement with Science & Technology provides resources for researchers engaging with the public (AAAS 2013). Professional organizations, disciplinary societies, and academies (including national young academies) have expertise that could be utilized to develop new and specialized resources in this area.

Box 10-4. Discussion Scenario: The News Release

You are a member of a chemical engineering research team whose paper has been accepted in a prestigious journal. Your institution's media relations office has drafted a news release about the paper. While the release does not contain any factual inaccuracies, some of the tone and language imply that the findings of your team will have a significant near-term impact on a particular industry. This is possible but not likely.

How would you communicate your concerns to other members of the research team and to the media relations office?

Today, new tools such as blogs, videos, and tweets are providing innovative ways for researchers to engage with the public. Engaging with colleagues and the broader public in the new social media environment also involves significant challenges and possible pitfalls, such as how to deal with intransigence, incivility, and personal attacks in communicating about issues with significant policy implications, such as climate change.

References

Chapter 1: Responsible Conduct of Research and the Global Context: An Overview

ALLEA (All European Academies) *Listing of online publications on research integrity.* Available at: www.allea.org/Pages/ALL/24/581.bGFuZz1FTkc.html. Accessed May 24, 2013.

Ethics CORE. University of Illinois. Available at: nationalethics center.org. Accessed May 24 , 2013.

IAC (InterAcademy Council) and IAP—The Global Network of Science Academies. 2012. *Responsible Conduct in the Global Research Enterprise: A Policy Report.* Amsterdam: IAC and IAP. Available at: www.interacademycouncil.net/24026/Global Report.aspx. Accessed 10 June 2013.

ICSU (International Council for Science). 2014. *Freedom, Responsibility and Universality of Science.* Available at: http://www.icsu .org/publications/cfrs/freedom-responsibility-and-universality -of-science-booklet-2014. Accessed January 28, 2015.

Glimcher, P. 2003. *Decisions Uncertainty and the Brain: The Science of Neuroeconomics.* Cambridge, MA: MIT Press.

Macrina, F. L. 2014. *Scientific Integrity,* 4th ed. Washington, DC: ASM Press.

NAS-NAE-IOM (National Academy of Sciences, National Academy of Engineering, Institute of Medicine). 2009. *On Being a Scientist: A Guide to Responsible Conduct in Research,* 3rd ed. Washington, DC: National Academies Press. Available at: http://www.nap.edu/openbook.php?record_id=12192. Accessed January 31, 2014.

NSB (National Science Board). 2012. *Science and Engineering Indicators.* Arlington, VA: National Science Foundation. Available at: http://www.nsf.gov/statistics/seind12/. Accessed January 31, 2014.

Nussinov, R., and C. Alemán. 2006. Nanobiology: From physics and engineering to biology. *Physical Biology* 3(1).

Online Ethics Center for Engineering and Science (OEC). Available at: www.onlineethics.org. Accessed May 24, 2013.

Resources for Research Ethics Education. 2013. University of California, San Diego. Available at: research-ethics.net. Accessed May 24, 2013.

Chapter 2: Planning and Preparing for Research

AG-NHMRC-UA (Australian Government, National Health and Medical Research Council, Universities Australia). 2007. *Australian Code for the Responsible Conduct of Research. Canberra.* Available at: http://www.nhmrc.gov.au/_files_nhmrc /publications/attachments/r39.pdf. Accessed January 31. 2014.

Bass, S. A., J. C. Rutledge, E. B. Douglass, and W. Y. Carter. 2007. *The University as Mentor: Lessons Learned from UMBC Inclusiveness Initiatives.* Washington, DC: Council of Graduate Schools. Available at: http://www.cgsnet.org/ckfinder/userfiles/files /Paper_Series_UMBC.pdf. Accessed January 31, 2014.

BWF-HHMI (Burroughs Wellcome Fund and Howard Hughes Medical Institute). 2006. Making the Right Moves: A Practical Guide to Scientific Management for Postdocs and New Faculty, 2nd ed., Bonetta L, ed. Research Triangle Park, NC, and Chevy Chase, MD: BWF-HHMI. Available at: http://www .hhmi.org/sites/default/files/Educational%20Materials/Lab%20 Management/Making%20the%20Right%20Moves/moves2.pdf. Accessed January 31, 2014.

CAS (Chinese Academy of Sciences). 2007. *Statements on the Notion of Science.* Beijing: CAS.

CCA (Council of Canadian Academies). 2010. *Honesty, Accountability and Trust: Fostering Research Integrity in Canada, Report of the Expert Panel on Research Integrity.* Ottawa: CCA. Available at: http://www.scienceadvice.ca/uploads/eng/assessments %20and%20publications%20and%20news%20releases/research %20integrity/ri_report.pdf. Accessed January 31, 2014.

DFG (Deutsche Forschungsgemeinschaft). 2013. *Good Scientific Practice at German Higher Education Institutions.* Bonn: DFG. Available at: http://www.hrk.de/uploads/tx_szconvention /Empfehlung_GutewissenschaftlichePraxis_MV_14042013 _EN.pdf. Accessed February 11, 2015.

ESF (European Science Foundation). 2010. *Fostering Research Integrity in Europe: A Report by the Member Organization Forum on Research Integrity.* Strasbourg: ESF. Available at: http://www.esf.org/fileadmin/Public_documents /Publications/ResearchIntegrity_report.pdf. Accessed January 31, 2014.

ESF-ALLEA (European Science Foundation and ALL European Academies). 2011. *The European Code of Conduct for Research Integrity.* Strasbourg: ESF. Available at: http://www.nsf.gov/od /iia/ise/Code_Conduct_ResearchIntegrity.pdf. Accessed January 31, 2014.

Friesenhahn, I., and C. Beaudry. 2014. *The Global State of Young Scientists. Project Report and Recommendations.* Available at: http://www.globalyoungacademy.net/projects/glosys-1/glosys -publications. Accessed January 28, 2015.

GBAU (General Board of the Association of Universities). 2004. *Netherlands Code of Conduct for Scientific Practice: Principles of Good Scientific Teaching and Research.* Amsterdam: Association of Universities in the Netherlands. Available at: https:// www.unesco-ihe.org/sites/default/files/the_netherlands _code_of_conduct_for_scientific_practice.pdf. Accessed January 31, 2014.

IAS (Indian Academy of Sciences). 2005. *Scientific Values: Ethical Guidelines and Procedures.* Bangalore: IAS. Available at: http:// www.ias.ac.in/academy/sci_val/scival-report.pdf. Accessed January 31, 2014.

ICB (Irish Council for Bioethics). 2010. *Recommendations for Promoting Research Integrity.* Dublin: The Irish Council for Bioethics. Available at: http://www.dohc.ie/working_groups /Current/nacb/Recommendations_for_Promoting.pdf?direct =1. Accessed January 31, 2014.

JANU (The Japan Association of National Universities), JAPU (The Japan Association of Public Universities), FJPCUA

(Federation of Japanese Private Colleges and Universities Associations), SCJ (Science Council of Japan). 2014. Joint Statement for Enhancing the Integrity of Scientific Research. December 11. Available at: http://www.janu.jp/eng/files/20141211 -y-017_en.pdf. Accessed June 3, 2015.

Johnson, V. E. 2008. Statistical analysis of the National Institutes of Health peer review system. *Proceedings of the National Academy of Sciences* 105 (32) 11076–80. Available at: http:// www.ncbi.nlm.nih.gov/pmc/articles/PMC2488382/. Accessed January 31, 2014.

KVA (Royal Swedish Academy of Sciences). 2012. Unexpected Benefits. Stockholm: KVA. Available at: http://www.kva.se /Documents/Vetenskap_samhallet/unexpected_benefits_2012 .pdf. Accessed June 19, 2013.

Lee, A., C. Dennis, and P. Campbell. 2007. Nature's guide for mentors. *Nature* 447: 791–97. Available at: http://www.nature .com/nature/journal/v447/n7146/full/447791a.html. Accessed January 31, 2014.

NAS-NAE-IOM (National Academy of Sciences, National Academy of Engineering, Institute of Medicine). 1997. *Adviser, Teacher, Role Model, Friend: On Being a Mentor to Students in Science and Engineering.* Washington, DC: National Academies Press. Available at: http://www.nap.edu/openbook .php?record_id=5789&page=R1. Accessed January 31, 2014.

———. 2009. *On Being a Scientist: A Guide to Responsible Conduct in Research, 3rd ed.* Washington, DC: National Academies Press. Available at: http://www.nap.edu/openbook.php?record_id =12192. Accessed January 31, 2014.

National Pact for Women in MINT Careers (Nationaler Pakt Für Frauen in MINT-Berufen). 2013. Web page. Available at: http:// www.komm-mach-mint.de/MINT-Projekte2/(adr_id)/74883 /(event_id)/1247/(lid)/05/(ort)/Bonn/(if)/81%2C80. Accessed June 19, 2013.

NRC-IOM (National Research Council, Institute of Medicine). 2002. *Integrity in Scientific Research: Creating an Environment That Promotes Responsible Conduct.* Washington, DC: National Academies Press. Available at: http://www.nap.edu/catalog .php?record_id=10430. Accessed January 31, 2014.

NRMN (National Research Mentoring Network). 2015. Web page. Available at: https://ictr.wisc.edu/files/ICTR_Today_v7n5 _FINALa.pdf. Accessed February 5, 2015.

O'Carroll, C. 2009. International peer review improved Irish research rankings. *Nature* 460 (949). Available at: http://www .nature.com/nature/journal/v460/n7258/full/460949a.html. Accessed January 31, 2014.

Piguet, V., Y. Tokura, and K. Green. 2011. Systematic mentoring on three continents. *Journal of Investigative Dermatology* 131: 549–50. Available at: http://www.nature.com/jid/journal/v131 /n3/pdf/jid2010424a.pdf. Accessed January 31, 2014.

PRL (Physical Review Letters). 2014. Editorial Policies and Practices. December. Available at: http://journals.aps.org/prl /authors/editorial-policies-practices. Accessed June 3, 2015.

RIA (Royal Irish Academy). 2010. *Ensuring Integrity in Irish Research: A Discussion Document.* Dublin: Royal Irish Academy. Available at: http://www.interacademies.net/File.aspx?id =14686. Accessed January 31, 2014.

Roig, M. 2006. Avoiding *Plagiarism, Self-plagiarism, and Other Questionable Writing Practices: A Guide to Ethical Writing.* Revised version. Available at: http://www.cse.msu.edu/~alexliu /plagiarism.pdf. Accessed February 25, 2015.

SCJ (Science Council of Japan). 2006. *Code of Conduct for Scientists.* Available at http://www.scj.go.jp/ja/info/kohyo/pdf/kohyo-20 -s3e.pdf. Accessed August 10, 2012.

Steneck, N. H. 2007. *ORI Introduction to the Responsible Conduct of Research.* Washington, DC: U.S. Department of Health and Human Services. Available at: http://ori.hhs.gov/sites/default /files/rcrintro.pdf. Accessed January 31, 2014.

3rd WCRI (3rd World Conference on Research Integrity). 2013. *Montreal Statement on Research Integrity.* Available at: http:// www.wcri2013.org/doc-pdf/MontrealStatement.pdf. Accessed February 11, 2015.

Van Noorden, R. 2009. Italy outsources peer review to NIH. *Nature* 459 (900). Available at: http://www.nature.com/news/2009 /090617/full/459900a.html. Accessed January 31, 2014.

Viereck, G. S. 1929. What life means to Einstein. *Saturday Evening Post.* October 26. Available at: http://www.saturdayevening

post.com/wp-content/uploads/satevepost/what_life_means
_to_einstein.pdf. Accessed June 17, 2013.

Chapter 3: Preventing the Misuse of Research and Technology

Barash, J. R., and S. S. Arnon. 2013. A novel strain of *Clostridium botulinum* that produces type B and type H botulinum toxins. *Journal of Infectious Diseases* 209: 183–91. Available at: http://jid.oxfordjournals.org/content/early/2013/10/07/infdis.jit449.full.pdf+html. Accessed January 27, 2015.

Berg, P., D. Baltimore, S. Brenner, R. O. Roblin III, and M. F. Singer. 1975. Summary statement of the Asilomar conference on recombinant DNA molecules. *Science* (188) 991. June 6. Available at: http://www.ncbi.nlm.nih.gov/pmc/articles/PMC432675/pdf/pnas00049-0007.pdf. Accessed January 31, 2014.

Cyranoski, D. and S. Reardon. 2015. Embryo editing sparks epic debate. *Nature* 520 (7549): 593–94 (April 30).

Drenth, P. J. D. 2012. Dual use and biosecurity: The case of the avian flu H5N1. *Open Journal of Applied Sciences* 2: 123–27. Available at: http://www.scirp.org/journal/PaperInformation.aspx?paperID=23256#.UuvIyrTN7vY. Accessed January 31, 2014.

Flower, R. J. 2011. Trends in science and technology relevant to the BTWC: Highlights from a meeting in Beijing in 2010. Presentation to IAP—The Global Network of Science Academies: Trends in science and technology relevant to the BWC, side event during the Preparatory Committee of the Seventh Review Conference, Geneva, April 14.

Frederickson, D. S. 1991. Asilomar and recombinant DNA: The end of the beginning. In K. E. Hanna, ed. *Biomedical Politics.* Washington, DC: National Academy Press. Available at: http://www.nap.edu/openbook.php?record_id=1793&page=258. Accessed January 31, 2014.

Herfst, S., E. J. Schrauwen, M. Linster, S. Chutinimitkul, E. de Wit, V. J. Munster, E. M. Sorrell, et al. 2012. Airborne transmission of influenza A/H5N1 virus between ferrets. *Science* 336:1534–41. Available at: http://dx.doi.org/ 10.1126/science.1213362. Accessed January 27, 2015.

IAP (InterAcademy Panel on International Issues). 2005. *IAP Statement on Biosecurity.* Available at: http://www.inter academies.net/File.aspx?id=540. Accessed January 31, 2014.

Imai, M., T. Watanabe, M. Hatta, S. C. Das, M. Ozawa , K. Shinya, G. Zhong, et al.. 2012. Experimental adaptation of an influenza H5 HA confers respiratory droplet transmission to a reassortant H5 HA/H1N1 virus in ferrets. *Nature* 486: 420–28. Available at: http://dx.doi.org/10.1038/nature10831. Accessed January 27, 2015.

Imperiale, M. J., and A. Casadevall. 2014. Vagueness and costs of the pause on gain-of-function (GOF) experiments on pathogens with pandemic, potential, including influenza virus. *MBio* 5(6): e02292-14. doi:10.1128/mBio.02292-14. Available at: http://mbio.asm.org/content/5/6/e02292-14.full. Accessed January 27, 2015.

KNAW (Royal Netherlands Academy of Arts and Sciences). 2008. *A Code of Conduct for Biosecurity: Report by the Biosecurity Working Group.* Available at: http://www.fas.org/biosecurity /resource/documents/IAP%20-%20Biosecurity%20code%20of% 20conduct.pdf. Accessed January 31, 2014.

NRC (National Research Council). 2004. *Biotechnology Research in an Age of Terrorism.* Washington, DC: National Academies Press. Available at: http://www.nap.edu/openbook.php?record _id=10827. Accessed January 31, 2014.

———. 2006. *Globalization, Biosecurity and the Future of the Life Sciences.* Washington, DC: National Academies Press. Available at: http://www.nap.edu/openbook.php?record_id=11567. Accessed January 31, 2014.

———. 2011a. *Review of the Scientific Approaches Used During the FBI's Investigation of the* 2001 *Anthrax Letters.* Washington, DC: National Academies Press. Available at: http://www.nap.edu /catalog/13098/review-of-the-scientific-approaches-used -during-the-fbis-investigation-of-the-2001-anthrax-letters. Accessed February 25, 2015.

———. 2011b. *Challenges and Opportunities for Education About Dual Use Issues in the Life Sciences.* In Cooperation with IAP, the International Union of Biochemistry and Molecular Biology, the International Union of Microbiological Societies, and the Polish Academy of Sciences. Washington, DC: National

Academies Press. Available at: http://dels.nas.edu/resources
/static-assets/materials-based-on-reports/reports-in-brief
/Dual-Use-Education-Report-Brief-FINAL.pdf. Accessed
January 31, 2014.

———. 2011c. *Life Sciences and Related Fields: Trends Relevant to
the Biological Weapons Convention.* In Cooperation with the
Chinese Academy of Sciences, IAP, the International Union
of Biochemistry and Molecular Biology, and the International
Union of Microbiological Societies. Washington, DC: National
Academies Press. Available at: http://www.nap.edu/catalog
/13130/life-sciences-and-related-fields-trends-relevant-to-the
-biological. Accessed January 27, 2015.

———. 2011d. *Research in the Life Sciences with Dual Use Potential:
An International Faculty Development Project on Education
About the Responsible Conduct of Science.* In Cooperation with
Bibliotheca Alexandrina and TWAS, the academy of sciences
for the developing world. Washington, DC: National Academies Press. Available at: http://www.nap.edu/openbook.php
?record_id=13270&page=R. Accessed January 31, 2014.

———. 2013. *Developing Capacities for Teaching Responsible Science in
the MENA Region: Refashioning Scientific Dialogue.* Washington,
DC: National Academies Press. Available at: http://www.nap.edu
/catalog/18356/developing-capacities-for-teaching-responsible
-science-in-the-mena-region. Accessed January 29, 2015.

OSTP (Office of Science and Technology Policy). 2014. *Doing
Diligence to Assess the Risks and Benefits of Life Sciences Gain-of-
Function Research.* Available at: http://www.whitehouse.gov
/blog/2014/10/17/doing-diligence-assess-risks-and-benefits-life
-sciences-gain-function-research. Accessed February 3, 2015.

Relman, D. A. 2013. "Inconvenient truths" in the pursuit of
scientific knowledge and public health. Editorial commentary.
Journal of Infectious Diseases. doi:10.1093/infdis/jit529. Available at: http://jid.oxfordjournals.org/content/early/2013/10/07
/infdis.jit529.full.pdf+html (subscription required).

Science. 2012. H5N1 special section. *Science* 336: 1521–47. Available
at: http://www.sciencemag.org/content/336/6088.toc. Accessed
September 12, 2013.

SCI (Science Council of Japan). 2013. Codes of conduct for scientists. Revised version. WHO (World Health Organization).

2004. *Laboratory Biosafety Manual*, 3rd ed. Available at: http://www.scj.go.jp/en/report/code.html. Accessed 31 January 2014.

USG (U.S. Government). 2012. *United States Government Policy for Oversight of Life Sciences Dual Use Research of Concern*. Available at: http://www.phe.gov/s3/dualuse/Documents/us-policy-durc-032812.pdf. Accessed August 6, 2015.

———. 2014. *United States Government Policy for Institutional Oversight of Life Sciences Dual Use Research of Concern*. Available at: http://www.phe.gov/s3/dualuse/Documents/durc-policy.pdf. Accessed August 6, 2015.

WHO. 2006. *Biorisk Management: Laboratory Biosecurity Guidance.* WHO/CDS/EPR/2006.6. Geneva: WHO. Available at: http://www.who.int/csr/resources/publications/biosafety/WHO_CDS_EPR_2006_6.pdf. Accessed January 31, 2014.

WHO. 2013. Report of the WHO Informal Consultation on Dual Use Research of Concern,Geneva, Switzerland, February 26–28, 2013. Available at: http://www.who.int/csr/durc/durc_feb2013_full_mtg_report.pdf. Accessed January 31, 2014.

Chapter 4: Carrying Out Research

Azoulay, P., J. L. Furman, J. L. Krieger, and F. E. Murray. 2012. Retractions. NBER Working Paper No. 18499. Available at: http://www.nber.org/papers/w18499. Accessed May 25, 2013.

Balstad, R. 2012. Overview of scientific data policies. In K. Bailey-Mathae and P. F. Uhlir, eds. *The Case for International Sharing of Scientific Data: A Focus on Developing Countries: Proceedings of a Symposium.* Washington, DC: National Academies Press. Available at: http://www.nap.edu/catalog.php?record_id=17019. Accessed January 31, 2014.

Begley, C. G. 2013. Reproducibility: Six red flags for suspect work. *Nature* (497) 493–94. Available at: http://www.nature.com/nature/journal/v497/n7450/full/497433a.html. Accessed June 19, 2013.

CGS (Council of Graduate Schools). 2012. *Research and Scholarly Integrity in Graduate Education.* Washington, DC: Council of Graduate Schools. Available at: http://cgsnet.org/research-and-scholarly-integrity-graduate-education-comprehensive-approach-0. Accessed January 31, 2014.

CODATA. 2013. CODATA Web site. Available at: http://www
.codata.org. Accessed May 25, 2013.

Drenth, P. J. D. 2013. *Institutional Responses to Violations of Research Integrity.* Paper presented at COPE European Seminar on Publication ethics, London. March 22. Available at: http://pieterdrenth.wordpress.com/2013/05/20/institutional-responses-to-violations-of-research-integrity/. Accessed June 22, 2013.

Economist. 2013. Trouble at the lab. October 19. Available at: http://www.economist.com/news/briefing/21588057-scientists-think-science-self-correcting-alarming-degree-it-not-trouble. Accessed November 5, 2013.

GEOSS. 2013. GEOSS Web site. Available at: http://www.earth observations.org/index.shtml. Accessed May 25, 2013.

Gilbert, N. 2009. Science journals crack down on image manipulation. *Nature News.* October 9. Available at: http://www.nature.com/news/2009/091009/full/news.2009.991.html. Accessed April 19, 2013.

Goodstein, D. 2010. *On Fact and Fraud: Cautionary Tales from the Front Lines of Science.* Princeton, NJ: Princeton University Press.

IAC (InterAcademy Council). 2006. *Women for Science: An Advisory Report.* Amsterdam: InterAcademy Council. Available at: http://www.interacademycouncil.net/File.aspx?id=27110. Accessed January 31, 2014.

Levelt Committee, Noort Committee, and Drenth Committee. 2012. *Flawed Science: The Fraudulent Ppractices of Social Psychologist Diederik Stapel.* Available at: http://www.tilburg university.edu/nl/nieuws-en-agenda/finalreportLevelt.pdf. Accessed April 19, 2013.

Molloy, J. C. 2011. The open knowledge foundation: Open data means better science. *PLoS Biology* 9: e1001195 doi:10.1371/journal.pbio.1001195.

NAS-NAE-IOM. 2009. *Ensuring the Integrity, Accessibility, and Stewardship of Research Data in the Digital Age.* Washington, DC: The National Academies Press. Available at: http://www.nap.edu/catalog.php?record_id=12615#toc. Accessed January 31, 2014.

Nature Cell Biology. 2006. Beautification and fraud (editorial). *Nature Cell Biology* 8101–2. Available at: http://www.nature

.com/ncb/journal/v8/n2/full/ncb0206-101.html. Accessed January 31, 2014.

Nature . 2013. *Challenges in Irreproducible Research.* Available at: http://www.nature.com/nature/focus/reproducibility/index .html. Accessed February 18, 2015.

NPG (Nature Publishing Group). 2013. Image integrity and standards. Available at: http://www.nature.com/authors/policies /image.html. Accessed April 19, 2013.

NRC (National Research Council). 2003. *Sharing Publication-Related Data and Materials: Responsibilities of Authorship in the Life Sciences.* Washington, DC: National Academies Press. Available at: http://www.nap.edu/catalog.php?record_id =10613. Accessed June 18, 2013.

———. 2011. *Expanding Underrepresented Minority Participation: America's Science and Technology Talent at the Crossroads.* Washington, DC: National Academies Press. Available at: http://grants.nih.gov/training/minority_participation.pdf. Accessed January 31, 2014.

Nyong, A., F. Adesina, and B. Osman Elasha. 2007. The value of indigenous knowledge in climate change mitigation and adaptation strategies in the African Sahel. *Mitigation and Adaptation Strategies for Global Change.* 12: 787–97. DOI 10.1007 /s11027-007-9099-0

Prinz, F., T. Schlange and K. Asadullah. 2011. Believe it or not: How much can we rely on published data on potential drug targets? *Nature Reviews Drug Discovery* 10, 712 (September 2011) | doi:10.1038/nrd3439-c1.

Royal Netherlands Academy of Arts and Sciences. 2013. *Responsible Research Data Management and the Prevention of Scientific Misconduct.* Amsterdam. Available at: https://www.knaw.nl /en/news/publications/responsible-research-data-management -and-the-prevention-of-scientific-misconduct/@@download /pdf_file/20131009.pdf. Accessed February 11, 2015.

The Royal Society. 2012. *Science as an Open Enterprise.* London. Available at: http://royalsociety.org/uploadedFiles/Royal _Society_Content/policy/projects/sape/2012-06-20-SAOE.pdf. Accessed January 31, 2014.

University of Alabama at Birmingham. 2008. *Online Learning Tool for Research Integrity and Digital Imaging.* Available at: http://

ori.hhs.gov/education/products/RIandImages/default.html. Accessed April 19, 2013.

Van Noorden, R. 2014a. Funders punish open-access dodgers. *Nature* doi:10.1038/508161a. Available at: http://www.nature.com /news/funders-punish-open-access-dodgers-1.15007. Accessed February 19, 2015.

———. 2014b. Publicly questioned papers more likely to be retracted. *Nature* doi:10.1038/nature2014.14979. Available at: http://www .nature.com/news/publicly-questioned-papers-more-likely-to -be-retracted-1.14979. Accessed February 11, 2015.

Chapter 5: The Researcher's Responsibilities to Society

AAAS (American Association for the Advancement of Science). 2012. *Connecting Science, Engineering, Ethics and Human Rights: Beyond Human Subjects Research.* Workshop Summary. Washington, DC: AAAS. Available at: http://www.aaas.org/sites /default/files/migrate/uploads/Workshop-Intersection-Science -Ethics-Human-Rights.pdf. Accessed January 31, 2014.

Andrade-C., M. G. 2012. Estado del conocimiento de la bio- diversidad en Colombia y sus amenazas. Consideraciones para fortalecer la interacción ambiente-política (State of knowledge of biodiversity in Colombia and threats: Considerations to strengthen environment-policy interaction). Revista de la Ac- ademia Colombiana Ciencias. 35 (137) 491–507, ISSN 0370-3908. Available at: http://accefyn.org.co/revista/Vol_35/137/492-508 .pdf. Accessed January 31, 2014.

———. 2013. Personal Web site: https://sites.google.com/site /mgandradec/. Accessed June 22, 2013.

Andrade-C., M. G., J. Betancur, E. Forero, J. Lynch, F. G. Stiles, and A. Prieto-C. 2012. Marco técnico y operativo para la con- strucción de la Estrategia del INB (Technical and operational framework for the construction of the BNI). In A. M. Suárez- Mayorga and J. C. Bello, ed., *Esquema conceptual y operativo para el desarrollo de la Enibio* (Conceptual and operational scheme for the development of the Enibio). Instituto de Inves- tigación de Recursos Biológicos Alexander von Humboldt y Ministerio de Ambiente y Desarrollo Sostenible (Alexander von Humboldt Research Institute for Biological Resources and Ministry of Environment and Sustainable Development). Bogotá, D.C.

Arora, S., A. L. Porter, J. Youtie and P. Shapira. 2012. *Capturing New Developments in Nanotechnology Scientific Output: A Search Strategy for Publication Records.* STIP White Paper. Atlanta, GA: Georgia Institute of Technology.

CBD (Convention on Biological Diversity). 2014. *Global Biodiversity Outlook 4. A Mid-term Assessment of Progress towards the Implementation of the Strategic Plan for Biodiversity 2011–2020.* Montreal, Quebec, Canada: Secretariat of the Convention on Biological Diversity. Available at: http://www.cbd.int/gbo /gbo4/publication/gbo4-en-lr.pdf. Accessed January 29, 2015.

Chen H. C., M. Dang, and M. C. Roco. 2010. Updated nanotechnology indicators, January 2010. Addendum to H. C. Chen and M. C. Roco, eds. 2008. *Mapping Nanotechnology Innovations and Knowledge: Global, Longitudinal Patent and Literature Analysis.* New York: Springer. Available at: http://www.springer.com /business+%26+management/business+information+systems /book/978-0-387-71619-0. Accessed January 31, 2014.

CIOMS (Council for International Organizations of Medical Sciences). 2002. *International Guidelines for Biomedical Research Involving Human Subjects.* Available at: http://www.cioms.ch /publications/guidelines/guidelines_nov_2002_blurb.htm. Accessed January 3, 2013.

COE-DH-BIO (Council of Europe, Committee on Bioethics). 1997. Convention for the Protection of Human Rights and Dignity of the Human Being with Regard to the Application of Biology and Medicine: Convention on Human Rights and Biomedicine (Oviedo Convention). Available at: conventions.coe .int/Treaty/en/Treaties/Html/164.htm. Accessed June 6, 2015.

———. 2003. Recommendation Rec(2003)10 of the Committee of Ministers to member states on xenotransplantation. Available at: wcd.coe.int/ViewDoc.jsp?id=45827. Accessed June 6, 2015.

———. 2006. Recommendation Rec(2006)4 of the Committee of Ministers to member states on research on biological materials of human origin. Available at: wcd.coe.int/ViewDoc .jsp?id=977859. Accessed June 6, 2015.

———. 2008. *Additional Protocol to the Convention on Human Rights and Biomedicine, Concerning Genetic Testing for Health Purposes.* Available at: conventions.coe.int/Treaty/en/Treaties/Html/203 .htm. Accessed June 6, 2015.

DHHS-OIG (Department of Health and Human Services, Office of the Inspector General). 2010. *Challenges to FDA's Ability to Monitor and Inspect Foreign Clinical Trials.* Washington, DC: DHHS. Available at: http://oig.hhs.gov/oei/reports/oei-01-08 -00510.pdf. Accessed January 31, 2014.

EC (European Commission). 2015. Protection of Personal Information. Web page: http://ec.europa.eu/justice/data-protection /index_en.htm. Accessed January 25, 2015.

El Espectador. 2013. Menos trámites para investigar en Colombia (Less bureaucracy for research in Colombia). July 2. Available at: http://www.elespectador.com/noticias/actualidad/vivir /articulo-431297-menos-tramites-investigar-colombia. Accessed July 9, 2013.

Fernández, F. 2011. The greatest impediment to the study of bio-diversity in Columbia. *Caldasia* 33(2):iii–v. Available at: http:// www.ciencias.unal.edu.co/unciencias/data-file/user_16/file /caldasia/3302/cld3302_ii.pdf. Accessed January 31, 2014.

Fog, L. 2011. Colombia to commercialize its biodiversity. *SciDev. Net.* July 6. Available at: http://www.scidev.net/global/bio diversity/news/colombia-to-commercialise-its-biodiversity .html. Accessed January 31, 2014.

Gertz, R. 2004. An analysis of the Icelandic Supreme Court judgement on the Health Sector Database Act. *SCRIPTed* (1:2). Available at: http://www2.law.ed.ac.uk/ahrc/script-ed/issue2 /iceland.asp. Accessed January 31, 2014.

Government of Colombia. 2009. Act 1333 of 2009, Environmental Penalty Procedures and Other Provisions (LEY 1333 DE 2009, Por la cual se establece el procedimiento sancionatorio ambi-ental y se dictan otras disposiciones). Official Gazette (47,417). 21 July. Available at: http://www.secretariasenado.gov.co /senado/basedoc/ley/2009/ley_1333_2009.html.. Accessed July 23, 2013.

———. 2013a. Decreto 1375 de 2013 (Decree 1375 of 2013). Available at: https://www.minambiente.gov.co/images/normativa /decretos/2013/dec_1375_2013.pdf. Accessed 9 April 2015.

———. 2013b. Decreto 1376 de 2013 (Decree 1376 of 2013). Available at: https://www.minambiente.gov.co/images/normativa /decretos/2013/dec_1376_2013.pdf. Accessed 9 April 2015.

Horizon 2020 Programme. *Call for Developing Governance for the Advancement of Responsible Research and Innovation: Reducing the Risk of Exporting Non Ethical Practices to Third Countries.* Available at: http://ec.europa.eu/research/participants/portal /desktop/en/opportunities/h2020/topics/2406-garri-6-2014 .html. Accessed February 18, 2015.

IHVN (Institute of Human Virology Nigeria). 2004. Web page, Available at: http://ihvnigeria.org/ihvnweb/webnew/index.php /about-ihvn/introduction.html. Accessed February 13, 2015.

Klochikhin, E. A., and P. Shapira. 2012. Engineering Small Worlds in a Big Society: Assessing the Early Impacts of Nanotechnology in China. *Review of Policy Research* 29(6) 752–75. November. Available at: http://onlinelibrary.wiley.com/doi/10.1111 /j.1541-1338.2012.00596.x/full. Accessed June 22, 2013.

Matthews, K. 2007. *Stem Cell Research: A Science and Policy Overview.* Materials produced for the conference series "Stem Cells: Saving Lives or Crossing Lines" sponsored by the James A. Baker III Institute for Public Policy at Rice University.

Ministry of Environment and Sustainable Development (El Ministerio de Ambiente y Desarrollo Sostenible). 2013. Ministry Web site. Available at: http://www.minambiente.gov.co /contenido/contenido.aspx?conID=1660&catID=609. Accessed June 22, 2013.

Ministry of Science and Technology and the Ministry of Health, People's Republic of China. 2003. *Ethical Guiding Principles on Human Embryonic Stem Cell Research.* December 24.

NAS-NAE-IOM (National Academy of Sciences, National Academy of Engineering, Institute of Medicine). 2009. *On Being a Scientist: A Guide to Responsible Conduct in Research, 3rd Edition.* Washington, DC: National Academies Press. Available at: http://www.nap.edu/catalog.php?record_id=12192. Accessed September 15, 2013.

NRC (National Research Council). 2010. *Promoting Chemical Laboratory Safety and Security in Developing Countries.* Washington, DC: National Academies Press. Available at: http://www

.nap.edu/catalog/12857/promoting-chemical-laboratory-safety
-and-security-in-developing-countries. Accessed February 4,
2015.

———. 2011a. *Guide for the Care and Use of Laboratory Animals:
Eighth Edition.* Washington, DC: National Academies Press.
Available at: http://grants.nih.gov/grants/olaw/Guide-for-the
-care-and-use-of-laboratory-animals.pdf. Accessed January 31,
2014.

———. 2011b. *Prudent Practices in the Laboratory: Handling and Management of Chemical Hazards, Updated Version.* Washington, DC:
National Academies Press. Available at: http://www.nap.edu
/catalog.php?record_id=12654. Accessed September 15, 2013.

———. 2015. *Climate Intervention: Reflecting Sunlight to Cool Earth.*
Washington, DC: National Academies Press. Available at:
http://www.nap.edu/catalog/18988/climate-intervention
-reflecting-sunlight-to-cool-earth. Accessed July 31, 2015.

Nuremburg Code. 1949. In Trials of War Criminals before Nuremburg Military Tribunals, Washington, DC: US Government
Printing Office. Available at: http://history.nih.gov/research
/downloads/nuremberg.pdf. Accessed January 28, 2015.

Obama, B. 2009. *Presidential Remarks on the Signing of Stem Cell
Executive Order and Scientific Integrity Presidential Memoran-
dum.* March 9. Available at: http://www.whitehouse.gov
/the_press_office/Remarks-of-the-President-As-Prepared-for
-Delivery-Signing-of-Stem-Cell-Executive-Order-and
-Scientific-Integrity-Presidential-Memorandum Accessed
September 16, 2013.

Qiu, J. 2012. Nano-safety studies urged in China. *Nature* 489: 350.
September 20. Available at: http://www.nature.com/news/nano
-safety-studies-urged-in-china-1.11437. Accessed July 10, 2013.

The Royal Society. 2009. *Geoengineering the Climate: Science, Governance and Uncertainty.* London: The Royal Society. Available
at: http://royalsociety.org/uploadedFiles/Royal_Society
_Content/policy/publications/2009/8693.pdf. Accessed
January 31, 2014.

SSP (Satellite Sentinel Project). 2013. SSP Web site. Available at:
http://www.satsentinel.org/. Accessed January 4, 2013.

UNESCO (United Nations Educational, Scientific and Cultural Organization). 2005. *Universal Declaration on Bioethics and Human*

Rights. Paris: UNESCO. Available at: http://www.unesco.org /new/en/social-and-human-sciences/themes/bioethics/bio ethics-and-human-rights/. Accessed January 31, 2014.

———. 2008. *Bioethics Core Curriculum.* Paris: UNESCO. Available at: http://unesdoc.unesco.org/images/0016/001636/163613e.pdf. Accessed January 31, 2014.

WMA (The World Medical Association). 2008. *The World Medical Association's Declaration of Helsinki.* Available at: http://www .wma.net/en/30publications/10policies/b3/17c.pdf. Accessed January 28, 2015.

Chapter 6: Preventing and Addressing Irresponsible Practices

AAAS (American Association for the Advancement of Science). 2001. *Should There Be an Oath for Scientists and Engineers?* http://www.aaas.org/page/should-there-be-oath-scientists-and -engineers. Accessed February 18, 2015.

Antes, A. L., S. T. Murphy, E. P. Waples, M. D. Mumford, R. P. Brown, S. Connelly, and L. D. Devenport. 2009. A meta-analysis of ethics instruction effectiveness in the sciences. *Ethics & Behavior, 19*(5), 379–402. Available at: http://www.ncbi.nlm.nih .gov/pmc/articles/PMC2762211/. Accessed January 31, 2014.

CCA (Council of Canadian Academies). 2010. *Honesty, Accountability and Trust: Fostering Research Integrity in Canada, Report of the Expert Panel on Research Integrity.* Ottawa: CCA. Available at: http://www.scienceadvice.ca/uploads/eng/assessments%20 and%20publications%20and%20news%20releases/research%20 integrity/ri_report.pdf. Accessed January 31, 2014.

Cong, Y. 2013. Research Integrity in China: Perspective from biomedical research integrity education. Presentation at the 3rd World Conference on Research Integrity, Montreal. May 6.

Cressey, D. 2007. Hippocratic oath for scientists. *Nature Blog.* Available at: http://blogs.nature.com/news/2007/09/hippocratic _oath_for_scientist.html. Accessed February 18, 2015.

Drenth, P. J. D. 2015. Institutional Dealing with Scientific Misconduct. ERUDITIO 1:6 February–April (136–46) Available at: www.worldacademy.org/eruditio/files/issue-6/ej-i6-book.pdf. Accessed July 7, 2015.

European Science Foundation and ALL European Academies (ESF-ALLEA). 2011. *The European Code of Conduct for Research Integrity*. Strasbourg: ESF. Available at: http://www.nsf.gov/od /iia/ise/Code_Conduct_ResearchIntegrity.pdf. Accessed January 31, 2014.

Fanelli, D. 2009. How many scientists fabricate and falsify research? A systematic review and meta-analysis of survey data. *PLoS One* 4(5): e5738. Available at: http://www.plosone .org/article/info%3Adoi%2F10.1371%2Fjournal.pone.0005738. Accessed January 31, 2014.

Fang F. C., R. G. Steen, and A. Casadevall. 2012. Misconduct accounts for the majority of retracted scientific publications. *Proceedings of the National Academy of Sciences, USA* 109(42) 17028–33. Available at: http://www.pnas.org/content/early /2012/09/27/1212247109.full.pdf+html. Accessed January 31, 2014.

Global Science Forum (OECD). 2007. *Best Practices for Ensuring Scientific Integrity and Preventing Misconduct*. Available at: http://www.oecd.org/science/sci-tech/40188303.pdf. Accessed January 31, 2014.

Gunsalus, C. K. 1998. How to blow the whistle and still have a career afterwards. *Science and Engineering Ethics* 4: 51–64. Available at: http://poynter.indiana.edu/files/8713/4858/3595 /see-ckg1.pdf. Accessed February 18, 2015.

ICB (Irish Council for Bioethics). 2010. *Recommendations for Promoting Research Integrity*. Dublin: The Irish Council for Bioethics. Available at: http://www.dohc.ie/working_groups /Current/nacb/Recommendations_for_Promoting.pdf?direct =1. Accessed January 31, 2014.

New Scientist. 2012. Fraud fighter: Faked research is endemic in China. Available at: http://www.newscientist.com/article/mg 21628910.300-fraud-fighter-faked-research-is-endemic-in-china .html#.VNKgop3F8k0. Accessed February 4, 2015.

New York Times. 2014. *How Senator John Walsh Plagiarized a Final Paper*. Available at: http://www.nytimes.com/interactive/2014 /07/23/us/politics/john-walsh-final-paper-plagiarism.html. Accessed February 4, 2015.

NHMRC (National Health and Medical Research Council). 2007. *Australian Code for the Responsible Conduct of Research*.

Canberra, Australia. Available at: http://www.nhmrc.gov.au
/guidelines-publications/r39. Accessed February 17, 2015.

Retraction Watch. 2015. Web page. Available at: http://retraction
watch.com/. Accessed February 4, 2015.

Steen, R. G., A. Casadevall, and F. C. Fang. 2013. Why has the
number of scientific retractions increased? *PLOS ONE* doi:
10.1371/journal.pone.0068397. Available at: http://journals.plos
.org/plosone/article?id=10.1371/journal.pone.0068397. Accessed
February 11, 2015.

Steneck, N. H. 2004. *Introduction to the Responsible Conduct of
Research*, rev. ed. Washington, DC: U.S. Government Printing
Office. Available at: http://ori.hhs.gov/sites/default/files/rc
rintro.pdf. Accessed January 31, 2014.

Times Higher Education. 2013. *A Plague of Plagiarism at the Heart
of Politics*. Available at: http://www.timeshighereducation.co
.uk/features/a-plague-of-plagiarism-at-the-heart-of-politics
/2003781.article. Accessed February 4, 2015.

UUK (Universities UK). 2012. *Concordat to support research integrity*. London. Available at: http://www.universitiesuk.ac.uk
/highereducation/Pages/Theconcordattosupportresearch
integrity.aspx. Accessed May 24, 2013.

Van Noorden, R. 2011. The trouble with retractions. *Nature* (478)
26–28. Available at: http://www.nature.com/news/2011/111005
/full/478026a.html. Accessed February 11, 2015.

Wells, F. O., S. Lock, and M. J. G. Farthing. 2001. *Fraud and Misconduct in Biomedical Research*. London: BMJ Books.

Chapter 7: Aligning Incentives with Responsible Research

Casadevall, A., and F. C. Fang. 2012. Reforming science: Structural reforms. Infection and immunity. 80(3): 897–901. Available at: http://iai.asm.org/content/80/3/897.long. Accessed
September 24, 2015.

CGS (Council of Graduate Schools). 2012. *Research and Scholarly
Integrity in Graduate Education*. Washington, DC: Council of
Graduate Schools.

Franzoni, C., G. Scellato, and P. Stephan. 2011. Science policy.
Changing incentives to publish. *Science* 333: 702–3. Available

at: http://www2.gsu.edu/~ecopes/Science-2011-Franzoni-702-3
.pdf. Accessed February 17, 2015.

ICMJE (International Committee of Medical Journal Editors).
2013. ICMJE *Form for Disclosure of Potential Conflicts of Interest*. Available at: http://www.icmje.org/coi_disclosure.pdf.
Accessed July 10, 2013.

NAS-NAE-IOM (National Academy of Sciences, National Academy of Engineering, Institute of Medicine). 2009. *On Being a Scientist: A Guide to Responsible Conduct in Research,* 3rd ed.
Washington, DC: National Academies Press. Available at:
http://www.nap.edu/openbook.php?record_id=12192. Accessed
January 31, 2014.

NRC-IOM (National Research Council, Institute of Medicine).
2002. *Integrity in Scientific Research: Creating an Environment That Promotes Responsible Conduct.* Washington, DC: The National Academies Press. Available at: http://www.nap.edu
/catalog.php?record_id=10430. Accessed January 31, 2014.

Oreskes, N., and E. M. Conway. 2011. *Merchants of Doubt.* New
York: Bloomsbury Press.

Proctor, R. N. 2011. *Golden Holocaust: Origins of the Cigarette Catastrophe and the Case for Abolition.* Berkeley and Los Angeles:
University of California Press.

San Francisco Declaration on Research Assessment (DORA).
2014. Available at: http://www.ascb.org/dora/. Accessed February 11, 2015.

Steinbrook, R. 2008. The Gelsinger Case. In E. J. Emanuel,
C. Grady, R. A. Crouch, R. K. Lie, F. G. Miller, and D. Wendler,
eds. *The Oxford Textbook on Clinical Research Ethics.* New York:
Oxford University Press.

Stephan, P. 2012. Research efficiency: Perverse incentives. *Nature*
(484) 29–31. April 5. Available at: http://www.nature.com
/nature/journal/v484/n7392/full/484029a.html. Accessed
January 31, 2014.

Chapter 8: Reporting Research Results

arXiV. 2015. Data accessible at: http://arxiv.org/.

Bohannon, J. 2013. Who's afraid of peer review? *Science* (342) 60–
65. Available at: http://www.sciencemag.org/content/342/6154
/60.full. Accessed February 10, 2015.

Chung, Y. G., J. H. Eum, J. E. Lee, S. H. Shim, V. Sepilian, S. W. Hong, Y. Lee, et al. 2014. Human somatic cell nuclear transfer using adult cells. *Cell Stem Cell.* 14: 777–80.

Committee of Medical Journal Editors. 2014. *Recommendations for the Conduct, Reporting, Editing, and Publication of Scholarly Work in Medical Journals.* Available at: http://www.icmje.org /icmje-recommendations.pdf. Accessed January 29, 2015.

COPE (Committee on Publication Ethics). 2011. *Code of Conduct and Best Practice Guidelines for Journal Editors.* Available at: http://publicationethics.org/files/Code_of_conduct_for _journal_editors.pdf. Accessed January 31, 2014.

Cyranoski, D. 2013. Fallout from hailed cloning paper. *Nature* 497: 543–44. Available at: http://www.nature.com/news/fallout -from-hailed-cloning-paper-1.13078. Accessed June 20, 2013.

———. 2014. Research integrity: Cell-induced stress. *Nature* 511: 140–43. Available at: http://www.nature.com/news/research-integrity -cell-induced-stress-1.15507. Accessed February 20, 2015.

Fang F. C., R. G. Steen, and A. Casadevall. 2012. Misconduct accounts for the majority of retracted scientific publications. *Proceedings of the National Academy of Sciences, USA* 109(42) 17028-33. Available at: http://www.pnas.org/content/early /2012/09/27/1212247109.full.pdf+html. Accessed January 31, 2014.

The Guardian. 2015. What pushes scientists to lie? The disturbing but familiar story of Haruko Obokata. February18. Available at: http://www.theguardian.com/science/2015/feb/18/haruko -obokata-stap-cells-controversy-scientists-lie. Accessed February 19, 2015.

Gunsalus, C. K. 1998. How to blow the whistle and still have a career afterwards. *Science and Engineering Ethics* 4: 51–64. Available at: http://poynter.indiana.edu/files/8713/4858/3595 /see-ckg1.pdf. Accessed February 18, 2015.

ICMJE (International Committee of Medical Journal Editors). 2015. *Defining the Role of Authors and Contributors.* Available at: http://www.icmje.org/recommendations/browse/roles-and -responsibilities/defining-the-role-of-authors-and-contributors .html. Accessed September 24, 2015.

Kolata, G. 2013. Scientific articles accepted (personal checks, Too). *New York Times.* April 7. Available at: http://www .nytimes.com/2013/04/08/health/for-scientists-an-exploding

-world-of-pseudo-academia.html?pagewanted=all&_r=0.
Accessed January 31, 2014.

Obokata, H., Y. Sasai, N. Hitoshi, M. Kadota, M. Andrabi,
N. Takata, M. Tokoro, et al. 2014. RETRACTED: Bidirectional
developmental potential in reprogrammed cells with acquired
pluripotency. *Nature* 505: 676–80.

Obokata, H., T. Wakayama, Y. Sasai, K. Kojima, M. P. Vacanti,
N. Hitoshi, Y. Masayuki, and C. A. Vacanti. 2014. RETRACTED:
Stimulus-triggered fate conversion of somatic cells into pluri-
potency. *Nature* 505: 641–47.

The Open Science Initiative Working Group (OSI). 2015. *Mapping
the Future of Scholarly Publishing. First Edition. National Science
Communication Institute.* Available at: http://nationalscience
.org/wp-content/uploads/2015/02/OSI-report-Feb-2015.pdf.
Accessed February 10, 2015.

Retraction Watch. 2015. Information accessible at: http://retraction
watch.com/.

Roig, M. 2010. Plagiarism and self-plagiarism: What every author
should know. *Biochem Med* 20: 295–300. Available at: http://
www.biochemia-medica.com/content/plagiarism-and-self
-plagiarism-what-every-author-should-know. Accessed
January 31, 2014.

Science. 2006. Special online collection: Hwang et al. controversy—
Committee report, response, and background. Available at:
www.sciencemag.org/site/feature/misc/webfeat/hwang2005/.
Accessed June 7, 2015.

Tachibana, M., P. Amato, M. Sparman, N. Marti Gutierrez,
R. Tippner-Hedges, H. Ma, E. Kang, et al. 2013a. Human
embryonic stem cells derived by somatic cell nuclear transfer.
Cell. 153: 1228–38. Available at: http://www.sciencedirect.com
/science/article/pii/S0092867413005710. Accessed February 20,
2015.

———. 2013b. Human embryonic stem cells derived by somatic cell
nuclear transfer. *Cell.* 154: 465–66. Available at: http://www
.sciencedirect.com/science/article/pii/S0092867413008246.
Accessed February 20, 2015.

Yamada, M. B. Johannensson, I. Sagi, L. C. Burnett, D. H. Kort,
R. W. Prosser, D. Paull, et al. 2014. Human oocytes reprogram
adult somatic nuclei of a type 1 diabetic to diploid pluripotent

stem cell. *Nature* 510: 533–36. Available at: http://www.nature
.com/nature/journal/v510/n7506/full/nature13287.html.
Accessed February 19, 2015.

Chapter 9: Benefits and Challenges of International Collaborations

Adams, J. 2013. Collaborations: The fourth age of research. *Nature*
497: 557–60. May 30. Available at: http://www.nature.com
/nature/journal/v497/n7451/full/497557a.html. Accessed
June 22, 2013.

CGIAR. 2013. CGIAR Web site. Available at: http://www.cgiar
.org/. Accessed July 9, 2013.

CoML (Census of Marine Life). 2010. Web site: http://www.coml
.org/.

Council of Graduate Schools. 2013. *Selected Resources on Research
Ethics Education in International Collaborations.* Available at:
http://www.cgsnet.org/selected-resources-research-ethics
-education-international-collaborations. Accessed June 22, 2013.

Danforth Center (Donald Danforth Plant Science Center). 2015.
Institute for International Crop Improvement Web site. Available at: https://www.danforthcenter.org/scientists-research
/research-institutes/institute-for-international-crop
-improvement. Accessed September 28, 2015.

ESF-ALLEA (European Science Foundation and ALL European
Academies). 2011. *The European Code of Conduct for Research
Integrity.* Strasbourg: ESF. Available at: http://www.nsf.gov/od
/iia/ise/Code_Conduct_ResearchIntegrity.pdf. Accessed
January 31, 2014.

Heitman, E., and S. Litewka. 2011. International perspectives
on plagiarism and considerations for teaching international
trainees. *Urologic Oncology* 29: 104–8. Available at: http://www
.ncbi.nlm.nih.gov/pmc/articles/PMC3038591/. Accessed
January 31, 2014.

KNAW (Royal Netherlands Academy of Arts and Sciences). 2014.
International Scientific Cooperation: Challenges and Predicaments; Options for Risk Assessment. Available at: www.knaw
.nl/en/news/publications/international-scientific-cooperation
-challenges-and-predicaments. Accessed June 7, 2015.

Mayer, T. 2013. Keynote talk at the Third World Conference on Research Integrity, Montréal. May 5.

Newberry, B., K. Austin, W. Lawson, G. Gorsuch, and T. Darwin. 2011. Acclimating international graduate students to professional engineering ethics. *Science and Engineering Ethics* 17(1):171–94. Available at: http://aln.coe.ttu.edu/ethics/document /SEE%20Abstract_Newberry%20et%20al.pdf/. Accessed January 31, 2014.

NRC (National Research Council). 2011. *Examining Core Elements of International Research Collaboration: Summary of a Workshop.* Washington, DC: The National Academies Press. Available at: http://www.nap.edu/catalog/13192/examining-core-elements -of-international-research-collaboration-summary-of-a. Accessed February 17, 2015.

———. 2014. *Culture Matters: International Research Collaboration in a Changing World (Summary of a Workshop).* Washington, DC: The National Academies Press. Available at: http://www.nap .edu/download.php?record_id=18849. Accessed February 17, 2015.

OECD Global Science Forum. 2007. Best Practices for Ensuring Scientific Integrity and Preventing Misconduct. Available at: http://www.oecd.org/science/sci-tech/40188303.pdf. Accessed January 31, 2014.

2nd World Conference on Research Integrity. 2010. *Singapore Statement on Research Integrity.* Available at: http://www .singaporestatement.org/. Accessed June 22, 2013.

Steneck, N. H. 2013. Global research integrity training. *Science* 340: 552–53. Available at: http://www.sciencemag.org/content /340/6132/552.full. Accessed January 31, 2014.

Steneck, N. H., and M. S. Anderson, eds. 2010. *International Research Collaboration: Much to Be Gained, Many Ways to Get in Trouble.* New York: Routledge.

3rd World Conference on Research Integrity. 2013. *Montreal Statement.* Available at: http://wcri2013.org/Montreal_Statement _e.shtml. Accessed June 22, 2013 (check for final).

Vasconcelos, S. M., N. H. Steneck, M. Anderson, H. Masuda, M. Palacios, J. C. Pinto, and M. M. Sorenson. 2012. The new geography of scientific collaborations. Changing patterns in the geography of science pose ethical challenges for collaborations between established and emerging scientific powers. *EMBO Rep.* May 1. 13(5): 404–7. Available at: http://www.ncbi.nlm.nih.gov/pmc/articles/PMC3343361/. Accessed January 31, 2014.

Chapter 10: Communicating with Policymakers and the Public

AAAS (American Association for the Advancement of Science). 2013. Web site of the Center for Public Engagement with Science & Technology. Available at: http://www.aaas.org/programs/centers/pe/. Accessed June 19, 2013.

Institute de France. Académie des sciences. Web page. Available at: http://www.academie-sciences.fr/en/pairing.htm. Accessed June 20, 2013.

Baron, N. 2010. *A Guide to Making your Science Matter: Escape from the Ivory Tower.* Washington, DC: Island Press.

Brownson, R. C., C. Royer, R. Ewing, and T. D. McBride. 2006. Researchers and policymakers: Travelers in parallel universes. *American Journal of Preventive Medicine.* 30(2): 164–72. Available at: https://uchastings.edu/academics/faculty/adjunct/barnes/classwebsite/docs/ResearchersandPolicyMakers-TravelersinParallelUniverses.pdf. Accessed January 31, 2014.

Cartlidge, E. 2012. Aftershocks in the courtroom. *Science* 338 (6104): 184–88. October 12. Available at: http://www.sciencemag.org/content/338/6104/184.full.pdf?sid=d89b2a90-6eed-431c-a543-b67fa32eaca0. Accessed June 20, 2013.

CFRDS (Committee on Freedom and Responsibility in Science). 2014. *Freedom, Responsibility and Universality of Science (booklet, 2014).* Available at: http://www.icsu.org/publications/cfrs/freedom-responsibility-and-universality-of-science-booklet-2014/CFRS-brochure-2014.pdf. Accessed: August 20, 2014.

COMPASS. 1999. COMPASS Web page. Available at: http://www.compassonline.org/. Accessed February 2, 2015.

Editors of The Lancet. 2010. Retraction—Ileal-lymphoid-nodular hyperplasia, non-specific colitis, and pervasive developmental disorder in children. 375:9713 (February 6) (445). Available at: www.thelancet.com/journals/lancet/article/PIIS0140-6736(10)60175-4/abstract. Accessed June 7, 2015.

Goldacre, B. 2009. *Bad Science*. London: Fourth estate.

IAC (InterAcademy Council). 2010. *Climate Change Assessments: Review of the Policies and Procedures of the Intergovernmental Panel on Climate Change*. Amsterdam: IAC. Available at: http://reviewipcc.interacademycouncil.net/. Accessed June 20, 2013.

IAP (IAP—The Global Network of Science Academies). 2006. *IAP Statement on the Teaching of Evolution*. June 21. Available at: http://www.interacademies.net/10878/13901.aspx. Accessed June 20, 2013.

ICSU (International Council for Science). 2014. *Freedom, Responsibility and Universality of Science*. Available at: http://www.icsu .org/publications/cfrs/freedom-responsibility-and-universality -of-science-booklet-2014. Accessed January 28, 2015.

IHRN (International Human Rights Network of Academies and Scholarly Societies). 2012. *Statement by the Executive Committee: Conviction of Italian Scientists in L'Aquila Trial*. Available at: http://www7.nationalacademies.org/humanrights/cs/groups /chrsite/documents/webpage/chr_073302.pdf. Accessed June 20, 2013.

IPCC (Intergovernmental Panel on Climate Change). 2013. IPCC web page. Available at: http://www.ipcc.ch/. Accessed June 20, 2013.

Nierenberg, C. 2015. Disneyland measles outbreak confirmed to be linked to low vaccination rates. *Scientific American*. March 17. Available at: http://www.scientificamerican.com/article /disneyland-measles-outbreak-confirmed-to-be-linked-to-low -vaccination-rates/. Accessed March 19, 2015.

New Scientist. 2012. Editorial: Italian earthquake case is no anti-science witch-hunt. *New Scientist*. October 23. Available at: http://www.newscientist.com/article/dn22416-italian-earth quake-case-is-no-antiscience-witchhunt.html. Accessed June 20, 2013.

Normile, D. 2011. In wake of Fukushima disaster, Japan's scientists ponder how to regain public trust. *Science Insider*. November 28. Available at: http://news.sciencemag.org/scienceinsider/2011/11 /in-wake-of-fukushima-disaster.html. Accessed June 20, 2013.

Nosengo, N. 2012. L'Aquila verdict row grows. *Nature* 491: 15–16. November 1. Available at: http://www.nature.com/news /l-aquila-verdict-row-grows-1.11683. Accessed June 20 2013.

Royal Society Pairing Scheme. 2013. Web page. Available at: https://royalsociety.org/training/pairing-scheme/. Accessed February 2, 2015.

Wakefield, A. J., S. H. Murch, A. Anthony, J. Linnell, D. M. Casson, M. Malik, M. Berelowitz, et al. 1998. RETRACTED: Ileal-lymphoid-nodular hyperplasia, non-specific colitis, and pervasive developmental disorder in children. *The Lancet.* 351(9103): 637-41 (February 28). Available at: www.thelancet .com/journals/lancet/article/PIIS0140-6736(97)11096-0/fulltext. Accessed June 7, 2015.

InterAcademy Partnership
Committee on Research Integrity

Indira NATH (Co-Chair), Former Head, Department of Biotechnology, All India Institute of Medical Sciences, and Former Raja RAMANNA Fellow and Emeritus Professor, National Institute of Pathology (ICMR), New Delhi, India

Ernst-Ludwig WINNACKER (Co-Chair), Emeritus Professor, University of Munich, Germany

Renfrew CHRISTIE, Former Dean of Research, University of the Western Cape, Bellville, South Africa

Pieter DRENTH, Former President, Royal Netherlands Academy of Arts and Sciences and Former President, ALL European Academies, Amsterdam, The Netherlands

Paula KIVIMAA, Senior Researcher, Finnish Environment Institute, Helsinki, Founding Member and Former Executive Committee Member of Global Young Academy, and Senior Research Fellow, Science Policy Research Unit, University of Sussex, Brighton

LI Zhenzhen, Director, Research Department of Policy for Science and Technology Development in the Institute of Policy and Management and Director, Research Center for Ethics of Science and Technology, Chinese Academy of Sciences, Beijing

José A. LOZANO, General Secretary, Colombian Academy of Exact, Physical and Natural Sciences, Bogotá, Colombia

Barbara SCHAAL, Dean of the Faculty of Arts & Sciences and Mary-Dell CHILTON Distinguished Professor of Biology, Washington University in St. Louis, USA

Project Staff

Tom ARRISON, Study Director
Anne MULLER, Program Coordinator, IAC Secretariat
Steve OLSON, Consultant Writer
Lida ANESTIDOU, Senior Staff Officer
Nina BOSTON, Senior Program Assistant
Patricia CABEZAS, Christine Mirzayan Fellow

Biographical Sketches
of Committee Members

Indira Nath (Cochair) is Former Head, Department of Biotechnology, All India Institute of Medical Sciences, and Former Raja RAMANNA Fellow and Emeritus Professor, National Institute of Pathology (ICMR), New Delhi, India. She received an MBBS from the All India Institute of Medical Sciences (AIIMS), New Delhi, and later served on the Faculty of AIIMS, making pioneering contributions to immunology research by her seminal work on cellular immune responses in human leprosy and a search for markers for viability of the leprosy bacillus which is not cultivable. She has also mentored many MBiotech, MD, and PhD students and made contributions to education, medical and science policies, and women scientists' issues. She was a member of the Scientific Advisory Committee to Cabinet, Foreign Secretary INSA (1995–97), council member (1992–94 and 1998–2006) and vice president (2001–2003) of the Indian Academy of Sciences, Bangalore, and chairperson, Women Scientists Programme, DST (2003). Numerous awards have been conferred on her, notably: Padmashri (1999), Chevalier Ordre National du Merite, France (2003), Silver Banner, Tuscany, Italy (2003), L'Oreal UNESCO Award for Women in Science (Asia Pacific) (2002), SS Bhatnagar Award (1983), and the Basanti Devi Amir Chand Award by ICMR (1994). She was elected fellow of the Indian National Science Academy, Delhi; National Academy of Sciences (India), Allahabad (1988); Indian Academy of Sciences, Bangalore (1990); National Academy of Medical Sciences (India) (1992); Royal

College of Pathology (1992); and the Academy of Sciences for the Developing World (TWAS) (1995). She was awarded a DSc (hc) in 2002 by Pierre and Marie Curie University, Paris, France.

Professor Ernst-Ludwig Winnacker (Cochair) is Emeritus Professor, University of Munich, Germany. He was previously secretary general of the Human Frontier Science Program Organization (HFSPO). He studied chemistry at the Swiss Federal Institute of Technology (ETH Zurich), where he obtained his PhD in 1968. After postdoctoral work at the University of California in Berkeley and the Karolinska Institute in Stockholm from 1968 to 1972, he became assistant and then DFG Visiting Professor at the Institute for Genetics, University of Cologne. In 1977 he was appointed associate professor at the Institute of Biochemistry at the Ludwig Maximilians University of Munich, where he was made full professor in 1980. From 1984 to 1997, he was director of the Laboratory of Molecular Biology at the University of Munich Gene Center. He served as president of the German Research Foundation (DFG) from 1998 to 2006. From 2003 to 2004 he also chaired the European Heads of Research Councils (EUROHORCs). He served as secretary general of the European Research Council (ERC) from 2007 to 2009. Professor Winnacker is a member of the U.S. National Academy of Sciences, Institute of Medicine, and of the German Academy of Sciences Leopoldina. His main fields of research are virus-cell interaction, the mechanisms of gene expression in higher cells, and prion diseases.

Professor Renfrew Christie is the former dean of research at the University of the Western Cape, South Africa. A specialist in the politics and economics of energy and in the history of science and technology, his Oxford doctorate treated the electrification of South Africa over 70 years. A whistle-blower for the African National Congress, on the apartheid nuclear weapons program, he was imprisoned for terrorism for seven and a half years in Pretoria. He co-founded the Macro Economic Research Group and the National Institute for Economic policy, which helped set South Africa's economy right after apartheid. He holds the Certificate of Commendation of the Chief of the South African Navy, for contributions to the democratic transformation of the South African Navy after apartheid. For 24 years he was a member of the Board of Trustees of South Africa's premier human rights law unit, the UWC Community Law Centre, and he chaired the board for nineteen years. His handwriting was on the second draft of the South African Bill of Rights. He is a defense force service commissioner, whose task is to advise the minister on the conditions of service of South Africa's troops. He has chaired the South African Commonwealth Scholarships Selection Committee for seventeen years. He has held visiting fellowships in the Woodrow Wilson International Center for Scholars, Washington DC; the Stiftung fur Wissenschaft und Politik, then in Ebenhausen; and the Indian Ocean Peace Centre, in Perth, Western Australia. He has had the privilege of addressing the Groupe Crises of the Institut de France on the Quai de Conti, Paris. He was a visiting professor of history in the University of Kentucky, Lexington, for the Spring Term of 2015. He attended both the Lisbon and Singapore World Conferences on Research Integrity. He is a signatory on the Singapore Statement on Research Integrity. He is a member of the Academy of Science of South Africa and a fellow of the Royal Society of South Africa.

Pieter J. D. Drenth studied psychology from 1952 to 1958 and received his PhD in 1960 at the VU University Amsterdam. With a Fulbright scholarship, he studied and worked in the United States (New York University and Standard Oil Co. of New Jersey) from 1960 to 1961. From 1962 to 1967 he was lecturer in test theory and statistics, and from 1967 to

2006 he was professor in test and scale theory and work and organizational psychology at the VU University Amsterdam. He was visiting professor at Washington University in St. Louis (1966) and the University of Washington, Seattle (1977). From 1982 to 1987 he was Rector Magnificus at the VU University Amsterdam, and from 1990 to 1996 he was president of the Royal Netherlands Academy of Arts and Sciences. From 2000 to 2006 he was president, and since 2006 has been honorary president, of ALL European Academies (ALLEA, the European federation of national academies of sciences and humanities). For his scientific work he received two honorary doctorates (Gent, 1981, and Paris Sorbonne, 1996). Her Majesty the Queen of the Netherlands conferred on him the knighthood in the order of the Netherlands' Lion (1990) and the commandership in the order of Oranje Nassau (1996).

Paula Kivimaa received her PhD in organizations and management and is a senior researcher at the Finnish Environment Institute, a government research organization in Finland. Since 2003 she has carried out research on the emergence of ecoinnovations in energy and forest sectors and on policy evaluation related to climate, energy, and innovation policies. Her current research focuses on innovations in energy and transport systems and on climate policy integration. Dr. Kivimaa obtained her PhD from Helsinki School of Economics in 2008. In 2009 she was an IAP-selected Young Scientist in the World Economic Forum Annual Meeting of the New Champions. In 2010 she was among the Young Scientists who established a global organization of early career scientists, Global Young Academy, and acted as an executive committee member during the first year of operation.

Professor Li Zhenzhen works as a research fellow in the Institute of Policy and Management, Chinese Academy of Sciences (IPM-CAS), where she serves as the director of the

Research Department of Policy for Science and Technology Development and the Research Section of Science, Technology and Society. In addition, she is the executive director of both the Research Support Center of Scientific Norms and Ethics as well as the Research Support Center of Scientific Popularization and Education, Academic Divisions of the Chinese Academy of Sciences. and the executive deputy editor in chief for the academic journal *Science and Society*. Her research interests mainly lie in the field of social studies of science, ethics of science and technology, and science and technology policy. In recent years, she has taken charge of major research projects funded by the National Natural Science Foundation of China, Ministry of Science and Technology of China, China Association for Science and Technology, and Chinese Academy of Sciences. In addition, she has been involved in several consultation projects associated with scientific affairs for government departments and civil society and has participated in drafting policy papers and reviewing law texts.

José A. Lozano received his PhD in geology from Columbia University in 1974. He is a retired professor of the National University of Colombia (1963–91), where he occupied several academic administrative positions and was a member of varied administrative academic committees. Professor Lozano is presently general secretary (elected) and executive secretary (appointed) of the Colombian Academy of Exact, Physical and Natural Sciences. He is a correspondent member of the Spanish Academy of Sciences, the focal point for Colombia of the Interamerican Network of Academies of Sciences (IANAS) Science Education Program, president of the Colombian Formation Environmental Net (Red Colombiana de Formación Ambiental), and secretary of the Professional Colombian Council of Geology. His interests encompass science education, capacity building, earth system science with emphasis in marine geology, and environmental sciences and policies. His previous positions include director of the Marine Research Institute, José Benito Vives de Andréis Marine and Coastal Research Institute (INVEMAR), Punta de Betín, Santa Marta (1979–81); adjunct professor, Earth Sciences and Resources Institute, University of South Carolina (1987–90); national correspondent of the IUGS Commission for Marine Geology (1982–90); chairman of the National Committee International Geosphere, Biosphere Programme (IGBP) (1993–2004); secretary of the Caribbean Scientific Union (CCC) (2005–7); and coordinator of the IANAS Science Education Program (2006–10).

Barbara Schaal is the Dean of the Faculty of Arts and Sciences, and the Mary Dell Chilton Distinguished Professor, Washington University in St. Louis. Schaal was born in Berlin, Germany and grew up in Chicago, IL, USA. She graduated from the University of Illinois, Chicago with a degree in biology and received a Ph.D. from Yale University. She is a plant evolutionary biologist who uses DNA sequences to understand evolutionary processes such as gene flow, geographical differentiation, and the domestication of crop species. Her current research focuses on the evolutionary genomics of rice. She currently serves as chair of the Division on Earth and Life Studies at the National Research Council and is a member of President Obama's Council of Advisors for Science and Technology. She has been president of the Botanical Society of America and the Society for the Study of Evolution and is an elected member of the US National Academy of Sciences and the American Academy of Arts and Sciences. She was appointed as a US science envoy by former Secretary of State Hillary Clinton. In February 2015 Schaal became the president-elect of the American Association of the Advancement of Science (AAAS).

Index

the interacademy partnership

The InterAcademy Partnership (IAP) is a new umbrella organization formed by the merging of three established inter-academy networks. As such it is governed by the leaders of these three networks, now called IAP for Science, IAP for Research, and IAP for Health. The leadership of the new umbrella organization also includes representatives of four regional networks–in Africa, the Asia/Pacific region, Europe, and the Americas. IAP currently has 130 member academies, which together reach governments that represent 95% of the world's population. The official launch of the InterAcademy Partnership is expected to take place in early 2016.

Since 1993, **IAP for Science**, previously known as the InterAcademy Panel, has harnessed the power of the world's scientific community to address global challenges and promote science-based sustainable development. IAP for Science brings together 107 member academies to advise the global public and decision-makers on the scientific aspects of critical global issues, such as sustainable development, climate change, biotechnology and global health. It also works to improve science education and scientific literacy in member countries.

Since 2000, **IAP for Research**, previously known as the InterAcademy Council, has mobilized the best scientists and engineers worldwide to provide high quality, in-depth advice to the United Nations and the broader global community on critical issues such as the importance of building scientific and technological capacity worldwide, a sustainable energy future, and African agriculture. IAP for Research has also presented a review of the processes used by the UN's Intergovernmental Panel on Climate Change (IPCC), and, most recently, set out a broad vision of scientific responsibility in the global research enterprise.

Established in 2000, **IAP for Health**, previously known as the InterAcademy Medical Panel, is a global network of more than 70 medical academies and academies of science and engineering with medical sections. It is committed to improving health world-wide, for example by strengthening the capacity of academies to provide evidence-based advice to governments on health and science policy, especially in relation to the social and environmental determinants of health and the rising threat of non-communicable diseases, and by supporting projects by member academies to strengthen health research and higher education in their countries. Its signature Young Physician Leaders program addresses sustainability of leadership to manage emerging challenges in health and bring about change.

More information available at: www.interacademies.org